Math
FOR CIVIL SERVICE TESTS

Math

FOR CIVIL SERVICE TESTS

Jessika Sobanski

NEW YORK

 Published in the United States by LearningExpress, LLC, New York.

Library of Congress Cataloging-in-Publication Data:
Sobanski, Jessika.
Math for civil service tests / Jessika Sobanski.—1st ed.
p. cm.
ISBN 1-57685-428-0 (pbk. : alk. paper)
1. Mathematics—Examinations, questions, etc. I. Title.

QA43 .S664 2002
510'.76—dc21

2002008106

Printed in the United States of America

9 8 7 6 5 4 3 2 1

First Edition

ISBN 1-57685-428-0

For more information or to place an order, contact LearningExpress at:
900 Broadway
Suite 604
New York, NY 10003

Or visit us at:
www.learnatest.com

Contents

About the Author

Jessika Sobanski is a math writer, teacher, and computer consultant. She is the author of *Visual Math* and *Math Builder*, and the coauthor of several other educational books. She lives in Long Island, New York.

Introduction

Choosing a career as a government employee can be very rewarding. But before you begin your job, you will find you must take a Civil Service exam. Civil Service exams require that candidates score well on all parts of the exam, but the math section can be especially daunting if it has been a long time since you have used your math skills. By making the commitment to practice for the math section of the Civil Service exam, you are promising yourself increased scores and marketability.

HOW TO USE THIS BOOK

Whether your exam is months away or weeks away, this book will help you prepare. You should carefully read this chapter and the next one, so you can grasp effective strategies and learn to budget your preparation time wisely. Chapter 2 presents a 30-day Study Plan and a 14-day Study Plan. You can decide which of these plans is right for you, or you can create a more personalized plan. Remember to stick as closely as you can to your plan. Always keep your end-goal in mind. If you study hard the first time, you will not have to take this exam again—ever! Use the exercises in this book to get a feel for the mathematics topics presented on the exam. Review them accordingly, take a practice exam, and then get ready to walk into the exam room with plenty of self-confidence!

But first, let's review some basic math strategies:

MATH STRATEGIES

The suggestions in this section are tried and true. You may use one or all of them. Or, you may decide to pick and choose the combination that works best for you.

- It's best not to work in your head! Use your test book or scratch paper to take notes, draw pictures, and calculate. Although you might think that you can solve math questions more quickly in your head, that's a good way to make mistakes. Instead, write out each step.
- Before you begin to make your calculations, read a math question in chunks rather than straight through from beginning to end. As you read each chunk, stop to think about what it means. Then make notes or draw a picture to represent that chunk of information.
- When you get to the actual question, circle it. This will keep you more focused as you solve the problem.
- Glance at the answer choices for clues. If they are fractions, you should do your work in fractions; if they are decimals, you should work in decimals, etc.
- Make a plan of attack to help you solve the problem. If a question stumps you, try one of the backdoor approaches explained in the next section. These are particularly useful for solving word problems. When you get your answer, reread the circled question to make sure you have answered it. This helps avoid the careless mistake of answering the wrong question.
- Check your work after you get an answer. Test takers get a false sense of security when they get an answer that matches one of the multiple-choice answers. It could be right, but you should always check your work. Remember to:
  - ✓ Ask yourself if your answer is reasonable and if it makes sense.
  - ✓ Plug your answer back into the problem to make sure the problem holds together.
  - ✓ Do the question a second time, but use a different method.
  - ✓ Approximate when appropriate. For example:
    $5.98 + $8.97 is a little less than $15. (Add: $6 + $9)
    .9876 × 5.0342 is close to 5. (Multiply: 1 × 5)
- Skip hard questions and come back to them later. Mark them in your test book so you can find them quickly.

BACKDOOR APPROACHES FOR ANSWERING QUESTIONS THAT PUZZLE YOU

Remember those dreaded word problems in high school? Many of them are actually easier to solve by backdoor approaches. The two techniques that follow are terrific ways to solve multiple-choice word problems that you don't know how to solve with a straightforward approach. The first technique, *nice numbers*, is useful when there are unknowns (like x) in the text of the word problem, making the problem too abstract for you. Nice numbers are numbers that are easy to work with, like multiples of ten, for example. The second technique, *working backwards*, presents a quick way to substitute numeric answer choices back into the problem to see which one works.

Nice Numbers

- When a question contains unknowns, like x, plug *nice numbers* in for the unknowns. A nice number makes calculations easier and makes sense in the problem.
- Read the question with the nice numbers in place. Then solve it.
- If the answer choices are all numbers, the choice that matches your answer is the right one.
- If the answer choices contain unknowns, substitute the same nice numbers into all the answer choices. The choice that matches your answer is the right one. If more than one answer matches, do the problem again with different nice numbers. You will only have to check the answer choices that have already matched.

Example: Judi went shopping with p dollars in her pocket. If the price of shirts was s shirts for d dollars, what is the maximum number of shirts Judi could buy with the money in her pocket?

a. psd

b. $\frac{ps}{d}$

c. $\frac{pd}{s}$

d. $\frac{ds}{p}$

To solve this problem, let's try these nice numbers: $p = \$100$; $s = 2$; $d = \$25$. Now reread it with the numbers in place:

Judi went shopping with $100 in her pocket. If the price of shirts was 2 shirts for $25, what is the maximum number of shirts Judi could buy with the money in her pocket? Since 2 shirts cost $25, that means that 4 shirts cost $50, and 8 shirts cost $100. So our answer is 8. Let's substitute the nice numbers into all 4 answers:

a. $100 \times 2 \times 25 = 5{,}000$

b. $\frac{100 \times 2}{25} = 8$

c. $\frac{100 \times 25}{2} = 1{,}250$

d. $\frac{25 \times 2}{100} = \frac{1}{2}$

The answer is **b** because it is the only one that matches our answer of 8.

WORKING BACKWARDS

You can frequently solve a word problem by plugging the answer choices back into the text of the problem to see which one fits all the facts stated in the problem. The process is faster than you think because you'll probably only have to substitute one or two answers to find the right one. This approach works only when

- all of the answer choices are numbers.
- you are asked to find a simple number, not a sum, product, difference, or ratio.

Here's what to do:

1. Look at all the answer choices and begin with the one in the middle of the range. For example, if the answers are 14, 8, 2, 20, and 25, begin by plugging 14 into the problem.
2. If your choice doesn't work, eliminate it. Determine if you need a bigger or smaller answer.
3. Plug in one of the remaining choices.
4. If none of the answers work, you may have made a careless error. Begin again or look for your mistake.

Example: Juan ate $\frac{1}{3}$ of the jellybeans. Maria then ate $\frac{3}{4}$ of the remaining jellybeans, which left 10 jellybeans. How many jellybeans were there to begin with?

a. 60
b. 80
c. 90
d. 120
e. 140

Starting with the middle answer, let's assume there were 90 jellybeans to begin with:

Since Juan ate $\frac{1}{3}$ of them, that means he ate 30 ($\frac{1}{3} \times 90 = 30$), leaving 60 of them ($90 - 30 = 60$). Maria then ate $\frac{3}{4}$ of the 60 jellybeans, or 45 of them ($\frac{3}{4} \times 60 = 45$). That leaves 15 jellybeans ($60 - 45 = 15$).

The problem states that there were 10 jellybeans left, and we wound up with 15 of them. That indicates that we started with too big a number. Thus, 90, 120, and 140 are all wrong! With only two choices left, let's use common sense to decide which one to try. The next lower answer is only a little smaller than 90 and may not be small enough. So, let's try 60:

Since Juan ate $\frac{1}{3}$ of them, that means he ate 20 ($\frac{1}{3} \times 60 = 20$), leaving 40 of them ($60 - 20 = 40$). Maria then ate $\frac{3}{4}$ of the 40 jellybeans, or 30 of them ($\frac{3}{4} \times 40 = 30$). That leaves 10 jellybeans ($40 - 30 = 10$).

Since the remainder is 10 jellybeans, the right answer is **a.**

KINDS OF CIVIL SERVICE JOBS

Civil Service jobs range from clerical work to forestry, from social work to cartography, from painting to nursing. The government workforce is diverse with possibilities like these:

- Accounting
- Administration
- Agriculture
- Biology
- Budgetary work
- Cartography
- Chemistry
- Claims work
- Clerical work
- Conservation
- Court work
- Custodial work
- Defense-related work
- Drafting
- Educational service
- Electric
- Engineering
- Finance
- Firefighting
- Health services
- Human services
- Labor
- Law enforcement
- Machinist work
- Nursing
- Painting
- Postal work
- Service work
- Social work
- Technical
- Treasury work
- Visa examination

EARNINGS AND ADVANCEMENT

The government is the number one employer in our country. Government jobs are secure, have great holiday and vacation schedules, offer health insurance, and provide paid training for employees. Benefits include: 10 paid holidays a year, 13 to 26 paid vacation days a year, 13 sick days a year, group life insurance, medical and dental benefits, and a government pension plan.

Civilian government employees are grouped by the type of work they do. This is called the *series*. The level of their relative positions (based on difficulty) is called the *grade*. Each grade progresses upwards through *steps*. The higher the step, the more money you make. Depending on your prior education, you may enter the government pay scale at different grades. For example, high school graduates may enter at GS-2, whereas junior college graduates may enter at GS-4. Following is the pay schedule for 2002:

U.S. Office Of Personnel Management
Salary Table 2002-GS
INCORPORATING A 3.60% GENERAL INCREASE
Effective January 2002
2002 General Schedule
Hourly (B)/Overtime (O) Rates by Grade and Step

GS	B/O	1	2	3	4	5	6	7	8	9	10
1	B	7.07	7.31	7.54	7.78	8.01	8.15	8.38	8.62	8.63	8.84
	O	10.61	10.97	11.31	11.67	12.02	12.23	12.57	12.93	12.95	13.26
2	B	7.95	8.14	8.4	8.63	8.72	8.98	9.23	9.49	9.75	10.00
	O	11.93	12.21	12.6	12.95	13.08	13.47	13.85	14.24	14.63	15.00
3	B	8.67	8.96	9.25	9.54	9.83	10.12	10.41	10.70	10.99	11.27
	O	13.01	13.44	13.88	14.31	14.75	15.18	15.62	16.05	16.49	16.91
4	B	9.74	10.06	10.39	10.71	11.03	11.36	11.68	12.01	12.33	12.66
	O	14.61	15.09	15.59	16.07	16.55	17.04	17.52	18.02	18.50	18.99
5	B	10.89	11.26	11.62	11.98	12.35	12.71	13.07	13.44	13.80	14.16
	O	16.34	16.89	17.43	17.97	18.53	19.07	19.61	20.16	20.70	21.24
6	B	12.14	12.55	12.95	13.36	13.76	14.17	14.57	14.98	15.38	15.79
	O	18.21	18.83	19.43	20.04	20.64	21.26	21.86	22.47	23.07	23.69
7	B	13.49	13.94	14.39	14.84	15.29	15.74	16.19	16.64	17.09	17.54
	O	20.24	20.91	21.59	22.26	22.94	23.61	24.29	24.96	25.64	26.31
8	B	14.95	15.44	15.94	16.44	16.94	17.44	17.94	18.43	18.93	19.43
	O	22.43	23.16	23.91	24.66	25.41	26.16	26.91	27.27	27.27	27.27
9	B	16.51	17.06	17.61	18.16	18.71	19.26	19.81	20.36	20.91	21.46
	O	24.77	25.59	26.42	27.24	27.27	27.27	27.27	27.27	27.27	27.27
10	B	18.18	18.78	19.39	20.00	20.60	21.21	21.82	22.42	23.03	23.63
	O	27.27	27.27	27.27	27.27	27.27	27.27	27.27	27.27	27.27	27.27
11	B	19.97	20.64	21.3	21.97	22.64	23.30	23.97	24.63	25.30	25.96
	O	27.27	27.27	27.27	27.27	27.27	27.27	27.27	27.27	27.27	27.27
12	B	23.94	24.74	25.53	26.33	27.13	27.93	28.72	29.52	30.32	31.12
	O	27.27	27.27	27.27	27.27	27.27	27.27	27.27	27.27	27.27	27.27
13	B	28.47	29.41	30.36	31.31	32.26	33.21	34.16	35.11	36.06	37.00
	O	27.27	27.27	27.27	27.27	27.27	27.27	27.27	27.27	27.27	27.27
14	B	33.64	34.76	35.88	37.00	38.12	39.25	40.37	41.49	42.61	43.73
	O	27.27	27.27	27.27	27.27	27.27	27.27	27.27	27.27	27.27	27.27
15	B	39.57	40.89	42.21	43.53	44.85	46.16	47.48	48.80	50.12	51.44
	O	27.27	27.27	27.27	27.27	27.27	27.27	27.27	27.27	27.27	27.27

Salary Table 2002-GS
2002 General Schedule
INCORPORATING A 3.60% GENERAL INCREASE
Effective January 2002
Annual Rates by Grade and Step

GS	1	2	3	4	5	6	7	8	9	10
1	14757	15249	15740	16228	16720	17009	17492	17981	18001	18456
2	16592	16985	17535	18001	18201	18736	19271	19806	20341	20876
3	18103	18706	19309	19912	20515	21118	21721	22324	22927	23530
4	20322	20999	21676	22353	23030	23707	24384	25061	25738	26415
5	22737	23495	24253	25011	25769	26527	27285	28043	28801	29559
6	25344	26189	27034	27879	28724	29569	30414	31259	32104	32949
7	28164	29103	30042	30981	31920	32859	33798	34737	35676	36615
8	31191	32231	33271	34311	35351	36391	37431	38471	39511	40551
9	34451	35599	36747	37895	39043	40191	41339	42487	43635	44783
10	37939	39204	40469	41734	42999	44264	45529	46794	48059	49324
11	41684	43073	44462	45851	47240	48629	50018	51407	52796	54185
12	49959	51624	53289	54954	56619	58284	59949	61614	63279	64944
13	59409	61389	63369	65349	67329	69309	71289	73269	75249	77229
14	70205	72545	74885	77225	79565	81905	84245	86585	88925	91265
15	82580	85333	88086	90839	93592	96345	99098	101851	104604	107357

The LearningExpress Test Preparation System

Taking any test can be tough. But, don't let the written test scare you! If you prepare ahead of time, you can achieve a top score. The LearningExpress Test Preparation System, developed exclusively for LearningExpress by leading test experts, gives you the discipline and attitude you need to be a winner.

First, the bad news: Getting ready for any test takes work! If you plan to obtain an entry-level Civil Service position, you will have to score well on your Civil Service Exam. This book focuses specifically on the math skills that you will be tested on. By sharpening these skills, you will take your first step toward achieving the career of your dreams. However, there are all sorts of pitfalls that can prevent you from doing your best on exams. Here are some obstacles that can stand in the way of your success.

- Being unfamiliar with the format of the exam
- Being paralyzed by test anxiety
- Leaving your preparation to the last minute
- Not preparing at all
- Not knowing vital test-taking skills, like:
  - how to pace yourself through the exam
  - how to use the process of elimination
  - when to guess
- Not being in tip-top mental and physical shape

- Forgetting to eat breakfast and having to take the test on an empty stomach
- Forgetting a sweater or jacket and shivering through the exam

What's the common denominator in all these test-taking pitfalls? One word: *control*. Who's in control, you or the exam?

Now the good news: The LearningExpress Test Preparation System puts *you* in control. In just nine easy-to-follow steps, you will learn everything you need to know to make sure that *you* are in charge of your preparation and your performance on the exam. *Other* test-takers may let the test get the better of them; *other* test-takers may be unprepared or out of shape, but not *you*. *You* will have taken all the steps you need to take for a passing score.

Here's how the LearningExpress Test Preparation System works: Nine easy steps lead you through everything you need to know and do to get ready to master your exam. Each of the steps listed below gives you tips and activities to help you prepare for any exam. It's important that you follow the advice and do the activities, or you won't be getting the full benefit of the system. Each step gives you an approximate time estimate.

Step 1. Get Information	30 minutes
Step 2. Conquer Test Anxiety	20 minutes
Step 3. Make a Plan	50 minutes
Step 4. Learn to Manage Your Time	10 minutes
Step 5. Learn to Use the Process of Elimination	20 minutes
Step 6. Know When to Guess	20 minutes
Step 7. Reach Your Peak Performance Zone	10 minutes
Step 8. Get Your Act Together	10 minutes
Step 9. Do it!	10 minutes
Total	**3 hours**

We estimate that working through the entire system will take you approximately three hours, though it's perfectly okay if you work faster or slower than the time estimates say. If you can take a whole afternoon or evening, you can work through the entire LearningExpress Test Preparation System in one sitting. Otherwise, you can break it up, and do just one or two steps a day for the next several days. It's up to you—remember, *you* are in control.

▶ STEP 1: GET INFORMATION

Time to complete: 30 minutes
Activities: Read Chapter 1, "Introduction."

If you haven't already done so, stop here and read Chapter 1 of this book. Here, you will learn how to use this book, review general math strategies, see an overview of the kinds of Civil Service jobs, and be presented with a discussion regarding earnings, advancement, and working conditions.

Knowledge is power. The first step in the LearningExpress Test Preparation System is finding out everything you can about the types of questions that will be asked on any math section of a Civil Service examination. Practicing and studying the exercises in this book will help prepare you for those tests. Mathematics topics that are tested include:

- Arithmetic, powers, and roots
- Fractions
- Decimals
- Number series and analogies
- Percents
- Word problems
- Charts, tables, and graphs
- Algebra
- Geometry and measurement

After completing the LearningExpress Test Preparation System, you will then begin to apply these test-taking strategies as you work through problem sets in the above topic areas (Chapters 3 through 10). You can see how well your training paid off in Chapters 11 and 12, where you will take two practice Civil Service examinations in math.

▶ STEP 2: CONQUER TEST ANXIETY

Time to complete: 20 minutes
Activity: Take the Test Stress Test

Having complete information about the exam is the first step in getting control of the exam. Next, you have to overcome one of the biggest obstacles to test success: test anxiety. Test anxiety not only impairs your performance on the exam itself, but it can even keep you from preparing! In Step 2, you will learn stress management techniques that will help you succeed on your exam. Learn these strategies now, and practice them as you work through the exams in this book, so they will be second nature to you by exam day.

Combating Test Anxiety

The first thing you need to know is that a little test anxiety is a good thing. Everyone gets nervous before a big exam—and if that nervousness motivates you to prepare thoroughly, so much the better. It's said that Sir Laurence Olivier, one of the foremost British actors of the last century, threw up before every performance. His stage fright didn't impair his performance; in fact, it probably gave him a little extra edge—just the kind of edge you need to do well, whether on a stage or in an exam room.

On the next page is the *Test Stress Test*. Stop here and answer the questions on that page, to find out whether your level of test anxiety is something you should worry about.

Stress Management Before the Test

If you feel your level of anxiety getting the best of you in the weeks before the test, here is what you need to do to bring the level down again:

- **Get prepared.** There is nothing like knowing what to expect. Being prepared will put you in control of test anxiety. That's why you are reading this book. Use it faithfully, and remind yourself that you are better prepared than most of the people taking the test.
- **Practice self-confidence.** A positive attitude is a great way to combat test anxiety. This is no time to be humble or shy. Stand in front of the mirror and say to your reflection, "I'm prepared. I'm full of self-confidence. I'm going to ace this test. I know I can do it." Say it into a tape recorder and play it back once a day. If you hear it often enough, you'll believe it.
- **Fight negative messages.** Every time someone starts telling you how hard the exam is or how it is almost impossible to get a high score, start telling them your self-confidence messages above. If the someone with the negative messages is you—telling yourself you don't do well on exams, you just can't do this—don't listen. Turn on your tape recorder and listen to your self-confidence messages.
- **Visualize.** Imagine yourself reporting for your first day on the job. Visualizing success can help make it happen—and it reminds you why you are preparing for the exam so diligently.
- **Exercise.** Physical activity helps calm your body down and focus your mind. Besides, being in good physical shape can actually help you do well on the exam. Go for a run, lift weights, go swimming—and do it regularly.

Stress Management on Test Day

There are several ways you can bring down your level of test anxiety on test day. To find a comfort level, practice these in the weeks before the test, and use the ones that work best for you.

- **Deep breathing.** Take a deep breath while you count to five. Hold it for a count of one, then let it out on a count of five. Repeat several times.
- **Move your body.** Try rolling your head in a circle. Rotate your shoulders. Shake your hands from the wrist. Many people find these movements very relaxing.
- **Visualize again.** Think of the place where you are most relaxed: lying on the beach in the sun, walking through the park, or whatever. Now, close your eyes and imagine you are actually there. If you practice in advance, you will find that you only need a few seconds of this exercise to experience a significant increase in your sense of well-being.

When anxiety threatens to overwhelm you right there during the exam, there are still things you can do to manage the stress level:

- **Repeat your self-confidence messages.** You should have them memorized by now. Say them quietly to yourself, and believe them!
- **Visualize one more time.** This time, visualize yourself moving smoothly and quickly through the test answering every question right and finishing just before time is up. Like most visualization techniques, this one works best if you have practiced it ahead of time.
- **Find an easy question.** Skim over the test until you find an easy question, and answer it. Getting even one circle filled in gets you into the test-taking groove.
- **Take a mental break.** Everyone loses concentration once in a while during a long test. It's normal, so you shouldn't worry about it. Instead, accept what has happened. Say to yourself, "Hey, I lost it there for a minute. My brain is taking a break." Put down your pencil, close your eyes, and do some deep breathing for a few seconds. Then you are ready to go back to work.

Try these techniques ahead of time, and see if they don't work for you!

TEST STRESS TEST

You only need to worry about test anxiety if it is extreme enough to impair your performance. The following questionnaire will provide a diagnosis of your level of test anxiety. In the blank before each statement, write the number that most accurately describes your experience.

0 = Never 1 = Once or twice 2 = Sometimes 3 = Often

_____ I have gotten so nervous before an exam that I simply put down the books and didn't study for it.

_____ I have experienced disabling physical symptoms such as vomiting and severe headaches because I was nervous about an exam.

_____ I have simply not showed up for an exam because I was scared to take it.

_____ I have experienced dizziness and disorientation while taking an exam.

_____ I have had trouble filling in the little circles because my hands were shaking too hard.

_____ I have failed an exam because I was too nervous to complete it.

_____ **Total:** Add up the numbers in the blanks above.

Your Test Stress Score

Here are the steps you should take, depending on your score. If you scored:

- Below 3, your level of test anxiety is nothing to worry about; it's probably just enough to give you the motivation to excel.
- Between 3 and 6, your test anxiety may be enough to impair your performance, and you should practice the stress management techniques listed in this section to try to bring your test anxiety down to manageable levels.
- Above 6, your level of test anxiety is a serious concern. In addition to practicing the stress management techniques listed in this section, you may want to seek additional, personal help. Call your local high school or community college and ask for the academic counselor. Tell the counselor that you have a level of test anxiety that sometimes keeps you from being able to take the exam. The counselor may be willing to help you or may suggest someone else you should talk to.

STEP 3: MAKE A PLAN

Time to complete: 50 minutes
Activity: Construct a study plan

Maybe the most important thing you can do to get control of yourself and your exam is to make a study plan. Too many people fail to prepare simply because they fail to plan. Spending hours on the day before the exam poring over sample test questions not only raises your level of test anxiety, it also is simply no substitute for careful preparation and practice.

Don't fall into the cram trap. Take control of your preparation time by mapping out a study schedule. If you are the kind of person who needs deadlines and assignments to motivate you for a project, here they are. If you are the kind of person who doesn't like to follow other people's plans, you can use the suggested schedules here to construct your own.

Even more important than making a plan is making a commitment. You can't review everything you need to know for a Civil Service test in one night. You have to set aside some time every day for study and practice. Try for at least 20 minutes a day. Twenty minutes daily will do you much more good than two hours on Saturday.

Don't put off your study until the day before the exam. Start now. A few minutes a day, with half an hour or more on weekends can make a big difference in your score.

If you have months before the exam, you are lucky. Don't put off your study until the week before the exam! Start now. Even ten minutes a day, with half an hour or more on weekends, can make a big difference in your score—and in your chances of making the grade you want!

Schedule A: The 30-day Plan

If you have at least a month before you take your test, you have plenty of time to prepare—as long as you don't waste it! If you have less than a month, turn to Schedule B.

TIME	PREPARATION
Day 1–2	Read Chapters 1 and 2 of this book. Also, skim over the written materials from any courses or training programs you may have taken, particularly noting: 1) the areas you expect to be emphasized on the exam and 2) the areas you don't remember well. On Day 4, concentrate on those areas.
Day 3	Read Chapter 3, Basic Arithmetic, and practice these basic skills by working through Questions 1–50.
Day 4	Read Chapter 4, Fractions. Work through Questions 1–10 and score yourself.
Day 5	Review any Chapter 4 concepts that you feel are necessary for you to brush up on. Work through Questions 11–30 and score yourself.
Day 6	Work through Questions 31–50 in Chapter 4. You should score yourself and make sure that you understand all of the concepts covered in this chapter.
Day 7	Read Chapter 5, Decimals, and work through Questions 1–10 and score yourself.
Day 8	Review any Chapter 5 concepts that you feel are necessary for you to brush up on. Work through Questions 11–30 and score yourself.
Day 9	Work through Questions 31–50 in Chapter 5. You should score yourself and make sure that you understand all of the concepts covered in this chapter.
Day 10	Read Chapter 6, Number Series and Analogies, and work through Questions 1–10 and score yourself.
Day 11	Review any Chapter 6 concepts that you feel are necessary for you to brush up on. Work through Questions 11–30 and score yourself.
Day 12	Work through Questions 31–49 in Chapter 6. You should score yourself and make sure that you understand all of the concepts covered in this chapter.
Day 13	Read Chapter 7, Percents, and work through Questions 1–10 and score yourself.
Day 14	Review any Chapter 7 concepts that you feel are necessary for you to brush up on. Work through Questions 11–30 and score yourself.
Day 15	Work through Questions 31–50 in Chapter 7. You should score yourself and make sure that you understand all of the concepts covered in this chapter.
Day 16	Read Chapter 8, Word Problems, and work through Questions 1–10. Score yourself.
Day 17	Review any Chapter 8 concepts that you feel are necessary for you to brush up on. Work through Questions 11–30 and score yourself.

Day 18	Work through Questions 31–50 in Chapter 8. You should score yourself and make sure that you understand all of the concepts covered in this chapter.
Day 19	Read Chapter 9, Charts, Tables, and Graphs, and work through Questions 1–10 and score yourself.
Day 20	Review any Chapter 9 concepts that you feel are necessary for you to brush up on. Work through Questions 11–30 and score yourself.
Day 21	Work through Questions 31–50 in Chapter 9. You should score yourself and make sure that you understand all of the concepts covered in this chapter.
Day 22	Read Chapter 10, Geometry and Measurement, and work through Questions 1–10. Score yourself.
Day 23	Review any Chapter 10 concepts that you feel are necessary for you to brush up on. Work through Questions 11–30 and score yourself.
Day 24	Work through Questions 31–50 in Chapter 10. You should score yourself and make sure that you understand all of the concepts covered in this chapter.
Day 25	In Chapter 11, take Practice Test 1. Score yourself and review any incorrect questions.
Day 26	Review any concepts that you feel are necessary for you to brush up on. Work through similar questions in the appropriate chapters.
Day 27	In Chapter 12, take Practice Test 2. Score yourself and review any incorrect questions.
Day 28	Review any concepts that you feel are necessary for you to brush up on. Work through similar questions in the appropriate chapters.
Day 29	Review the chapters that contained the topics that you were weak on during the Practice Exams.

Day before the exam: Relax. Do something unrelated to the exam and go to bed at a reasonable hour.

Schedule B: The 14-day plan

If you have two weeks or less before you take your exam, you may have your work cut out for you. Use this 14-day schedule to help you make the most of your time.

TIME	PREPARATION
Day 1	Read Chapters 1 and 2.
Day 2	Complete the entire Arithmetic chapter (Chapter 3) including the Practice Questions.
Day 3	Complete the entire Fractions chapter (Chapter 4) including the Practice Questions.
Day 4	Complete the entire Decimals chapter (Chapter 5) including the Practice Questions.
Day 5	Complete the entire Number Series/Analogies chapter (Chapter 6) including the Practice Questions.
Day 6	Complete the entire Percents chapter (Chapter 7) including the Practice Questions.
Day 7	Complete the entire Word Problems chapter (Chapter 8) including the Practice Questions.
Day 8	Complete the entire Charts, Tables, and Graphs chapter (Chapter 9) including the Practice Questions.
Day 9	Complete the entire Geometry and Measurement chapter (Chapter 10) including the Practice Questions.
Day 10	Complete Practice Test 1 (Chapter 11) and score yourself. Review all of the questions that you missed.
Day 11	Review any concepts that you feel are necessary for you to brush up on. Work through similar questions in the appropriate chapters.
Day 12	Complete Practice Test 2 (Chapter 12) and score yourself. Review all of the questions that you missed.
Day 13	Review any topics as indicated by the questions you missed on the Practice Test. Then look at the questions you missed again and make sure that you understand them.
Day before the exam: Relax. Do something unrelated to the exam and go to bed at a reasonable hour.	

STEP 4: LEARN TO MANAGE YOUR TIME

Time to complete: 10 minutes to read, many hours of practice!
Activities: Practice these strategies as you take the sample tests in this book

Steps 4, 5, and 6 of the LearningExpress Test Preparation System put you in charge of your exam by showing you test-taking strategies that work. Practice these strategies as you take the sample tests in this book, and then you will be ready to use them on test day.

First, take control of your time on the exam. Civil Service exams have a time limit, which may give you more than enough time to complete all the questions—or may not. It's a terrible feeling to hear the examiner say, "Five minutes left," when you are only three-quarters of the way through the test. Here are some tips to keep that from happening to *you.*

- **Follow directions**. If the directions are given orally, listen closely. If they are written on the exam booklet, read them carefully. Ask questions *before* the exam begins if there is anything you don't understand. If you are allowed to write in your exam booklet, write down the beginning time and the ending time of the exam.
- **Pace yourself**. Glance at your watch every few minutes, and compare the time to how far you've gotten in the test. When one-quarter of the time has elapsed, you should be a quarter of the way through the section, and so on. If you are falling behind, pick up the pace a bit.
- **Keep moving**. Don't waste time on one question. If you don't know the answer, skip the question and move on. Circle the number of the question in your test booklet in case you have time to come back to it later.
- **Keep track of your place on the answer sheet**. If you skip a question, make sure you skip on the answer sheet too. Check yourself every 5–10 questions to make sure the question number and the answer sheet number are still the same.
- **Don't rush.** Though you should keep moving, rushing won't help. Try to keep calm and work methodically and quickly.

STEP 5: LEARN TO USE THE PROCESS OF ELIMINATION

Time to complete: 20 minutes
Activity: Complete worksheet on Using the Process of Elimination

After time management, your next most important tool for taking control of your exam is using the process of elimination wisely. It's standard test-taking wisdom that you should always read all the answer choices before choosing your answer. This helps you find the right answer by eliminating wrong answer choices. And, sure enough, that standard of wisdom applies to your exam, too.

Choosing the Right Answer by Process of Elimination

As you read a question, you may find it helpful to underline important information or make some notes about what you are reading. When you get to the heart of the question, circle it and make sure you understand what it is asking. If you are not sure of what's being asked, you will never know whether you have chosen the right answer. What you do next depends on the type of question you are answering.

- If it's math, take a quick look at the answer choices for some clues. Sometimes this helps to put the question in a new perspective and makes it easier to answer. Then make a plan of attack to solve the problem.
- Otherwise, follow this simple process of elimination plan to manage your testing time as efficiently as possible: Read each answer choice and make a quick decision about what to do with it, marking your test book accordingly:
  - The answer seems reasonable; keep it. Put a ✔ next to the answer.
  - The answer is awful. Get rid of it. Put an **X** next to the answer.
  - You can't make up your mind about the answer, or you don't understand it. Keep it for now. Put a **?** next to it.

Whatever you do, don't waste time with any one answer choice. If you can't figure out what an answer choice means, don't worry about it. If it's the right answer, you will probably be able to eliminate all the others, and, if it's the wrong answer, another answer will probably strike you more obviously as the right answer.

If you haven't eliminated any answers at all, skip the question temporarily, but don't forget to mark the question so you can come back to it later if you have time. If the test has no penalty for wrong answers, and you are certain that you could never answer this question in a million years, pick an answer and move on!

If you have eliminated all but one answer, just reread the circled part of the question to make sure you are answering exactly what's asked. Mark your answer sheet and move on to the next question.

Here's what to do when you have eliminated some, but not all of the answer choices: Compare the remaining answers looking for similarities and differences, reasoning your way through these choices. Try to eliminate those choices that don't seem as strong to you. But DON'T eliminate an answer just because you don't understand it. You may even be able to use relevant information from other parts of the test. If you have narrowed it down to a single answer, check it against the circled question to be sure you have answered it. Then, mark your answer sheet and move on. If you are down to only two or three answer choices, you have improved your odds of getting the question right. Make an educated guess and move on. However, if you think you can do better with more time, mark the question as one to return to later.

If You're Penalized for Wrong Answers

You must know whether you will be penalized for wrong answers before you begin the test. If you don't, ask the proctor before the test begins. Whether you make a guess or not depends upon the penalty. Some standardized tests are scored in such a way that every wrong answer reduces your score by a fraction of a point, and these can really add up against you! Whatever the penalty, if you can eliminate enough choices to make the odds of answering the question better than the penalty for getting it wrong, make a guess. This is called educated guessing.

Let's imagine you are taking a test in which each answer has five choices and you are penalized $\frac{1}{4}$ of a point for each wrong answer. If you cannot eliminate any of the answer choices, you are better off leaving the answer blank because the odds of guessing correctly are one in five. However, if you can eliminate two of the choices as definitely wrong, the odds are now in your favor. You have a one in three chance of answering the question correctly. Fortunately, few tests are scored using such elaborate means, but if your test is one of them, know the penalties and calculate your odds before you take a guess on a question.

If You Finish Early

Use any time you have left to do the following:

- Go back to questions you marked to return to and try them again.
- Check your work on all the other questions. If you have a good reason for thinking a response is wrong, change it.
- Review your answer sheet. Make sure that you have put the answers in the right places and that you have marked only one answer for each question. (Most tests are scored in such a way that questions with more than one answer are marked wrong.)
- If you have erased an answer, make sure you have done a good job of it.
- Check for stray marks on your answer sheet that could distort your score.

Whatever you do, don't waste time when you have finished a test section. Make every second count by checking your work over and over again until time is called.

Try using your powers of elimination on the questions in the following worksheet called "Using the Process of Elimination." The answer explanations that follow show one possible way you might use the process to arrive at the right answer.

The process of elimination is your tool for the next step, which is knowing when to guess.

USING THE PROCESS OF ELIMINATION

Use the process of elimination to answer the following questions.

1. Ilsa is as old as Meghan will be in five years. The difference between Ed's age and Meghan's age is twice the difference between Ilsa's age and Meghan's age. Ed is 29. How old is Ilsa?

a. 4
b. 10
c. 19
d. 24

2. "All drivers of commercial vehicles must carry a valid commercial driver's license whenever operating a commercial vehicle." According to this sentence, which of the following people need NOT carry a commercial driver's license?
   a. a truck driver idling his engine while waiting to be directed to a loading dock
   b. a bus operator backing her bus out of the way of another bus in the bus lot
   c. a taxi driver driving his personal car to the grocery store
   d. a limousine driver taking the limousine to her home after dropping off her last passenger of the evening

3. Smoking tobacco has been linked to
   a. an increased risk of stroke and heart attack.
   b. all forms of respiratory disease.
   c. increasing mortality rates over the past ten years.
   d. juvenile delinquency.

4. Which of the following words is spelled correctly?
   a. incorrigible
   b. outragous
   c. domestickated
   d. understandible

Answers

Here are the answers, as well as some suggestions as to how you might have used the process of elimination to find them.

1. **d.** You should have eliminated choice **a** immediately. Ilsa can't be four years old if Meghan is going to be Ilsa's age in five years. The best way to eliminate other answer choices is to try plugging them in to the information given in the problem. For instance, for choice **b**, if Ilsa is 10, then Meghan must be 5. The difference in their ages is 5. The difference between Ed's age, 29, and Meghan's age, 5, is 24. Is 24 two times 5? No. Then choice **b** is wrong. You could eliminate choice **c** in the same way and be left with choice **d**.
2. **c.** Note the word *not* in the question, and go through the answers one by one. Is the truck driver in choice **a** "operating a commercial vehicle"? Yes, idling counts as "operating," so he needs to have a commercial driver's license. Likewise, the bus operator in choice **b** is operating a commercial vehicle; the question doesn't say the operator has to be on the street. The limo driver in choice **d** is operating a commercial vehicle, even if it doesn't have a passenger in it. However, the cabbie in choice **c** is not operating a commercial vehicle, but his own private car.
3. **a.**You could eliminate choice **b** simply because of the presence of the word all. Such absolutes hardly ever appear in correct answer choices. Choice **c** looks attractive until you think a little about what you know—aren't fewer people smoking these days, rather than more? So, how could smoking be responsible for a higher mortality rate? (If you didn't know that

mortality rate means the rate at which people die, you might keep this choice as a possibility, but you would still be able to eliminate two answers and have only two to choose from.) Choice **d** can't be proven, so you could eliminate that one, too. Now you are left with the correct choice, **a.**

4. **a.** How you used the process of elimination here depends on which words you recognized as being spelled incorrectly. If you knew that the correct spellings were outrageous, domesticated, and understandable, then you were home free. Surely, you knew that at least one of those words was wrong.

STEP 6: KNOW WHEN TO GUESS

Time to complete: 20 minutes
Activity: Complete Worksheet on Your Guessing Ability

Armed with the process of elimination, you are ready to take control of one of the big questions in test-taking: Should I guess? The first and main answer is, *Yes.* Some exams have what's called a "guessing penalty," in which a fraction of your wrong answers is subtracted from your right answers. Check with the administrators of your particular exam to see if this is the case. In many instances, the number of questions you answer correctly yields your raw score. So, you have nothing to lose and everything to gain by guessing.

The more complicated answer to the question, "Should I guess?" depends on you, your personality, and your "guessing intutition." There are two things you need to know about yourself before you go into the exam:

- Are you a risk-taker?
- Are you a good guesser?

You will have to decide about your risk-taking quotient on your own. To find out if you are a good guesser, complete the following worksheet called *Your Guessing Ability.* Frankly, even if you are a play-it-safe person with terrible intuition, you are still safe in guessing every time. The best thing would be if you could overcome your anxieties and go ahead and mark an answer. But you may want to have a sense of how good your intuition is before you go into the exam.

YOUR GUESSING ABILITY

The following are ten especially hard questions. You are not supposed to know the answers. Rather, this is an assessment of your ability to guess when you don't have a clue. Read each question carefully, just as if you did expect to answer it. If you have any knowledge at all of the subject of the question, use that knowledge to help you eliminate wrong answer choices. Circle the answer choice you believe to be correct.

1. September 7 is Independence Day in
- **a.** India.
- **b.** Costa Rica.
- **c.** Brazil.
- **d.** Australia.

2. Which of the following is the formula for determining the momentum of an object?
- **a.** $p = mv$
- **b.** $F = ma$
- **c.** $P = IV$
- **d.** $E = mc^2$

3. Because of the expansion of the universe, the stars and other celestial bodies are all moving away from each other. This phenomenon is known as
- **a.** Newton's first law.
- **b.** the big bang.
- **c.** gravitational collapse. ✓
- **d.** Hubble flow.

4. American author Gertrude Stein was born in
- **a.** 1713.
- **b.** 1830.
- **c.** 1874.
- **d.** 1901.

5. Which of the following is NOT one of the Five Classics attributed to Confucius?
- **a.** the I Ching
- **b.** the Book of Holiness
- **c.** the Spring and Autumn Annals ✓
- **d.** the Book of History

6. The religious and philosophical doctrine that holds that the universe is constantly in a struggle between good and evil is known as
   a. Pelagianism.
   b. Manichaeanism.
   c. neo-Hegelianism.
   d. Epicureanism.

7. The third Chief Justice of the U.S. Supreme Court was
   a. John Blair.
   b. William Cushing.
   c. James Wilson.
   d. John Jay.

8. Which of the following is the poisonous portion of a daffodil?
   a. the bulb
   b. the leaves
   c. the stem
   d. the flowers

9. The winner of the Masters golf tournament in 1953 was
   a. Sam Snead.
   b. Cary Middlecoff.
   c. Arnold Palmer.
   d. Ben Hogan.

10. The state with the highest per capita personal income in 1980 was
    a. Alaska.
    b. Connecticut.
    c. New York.
    d. Texas.

Answers

Check your answers against the correct answers below.

1. c.
2. a.
3. d.
4. c.
5. b.
6. b.
7. b.
8. a.
9. d.
10. a.

How Did You Do?

You may have simply gotten lucky and actually known the answer to one or two questions. In addition, your guessing was more successful if you were able to use the process of elimination on any of the questions. Maybe you didn't know who the third Chief Justice was (question 7), but you knew that John Jay was the first. In that case, you would have eliminated answer **d** and, therefore improved your odds of guessing right from one in four to one in three.

According to probability, you should get $2\frac{1}{2}$ answers correct, so getting either two or three right would be average. If you got four or more right, you may be a really terrific guesser. If you got one or none right, you may decide not to guess.

Keep in mind, though, that this is only a small sample. You should continue to keep track of your guessing ability as you work through the sample questions in this book. Circle the numbers of questions you guess; or, if you don't have time during the practice tests, go back afterward and try to remember which questions you guessed. Remember, on a test with four answer choices, your chances of getting a right answer is one in four. So, keep a separate "guessing" score for each exam. How many questions did you guess? How many did you get right? If the number you got right is at least one-fourth of the number of questions you guessed, you are at least an average guesser, maybe better—and you should always go ahead and guess on the real exam. If the number you got right is significantly lower than one-fourth of the number you guessed on, you should not guess on exams where there is a guessing penalty unless you can eliminate a wrong answer. If there's no guessing penalty, you would, frankly, be safe in guessing anyway, but maybe you would feel more comfortable if you guessed only selectively, when you can eliminate a wrong answer or at least have a good feeling about one of the answer choices.

STEP 7: REACH YOUR PEAK PERFORMANCE ZONE

Time to complete: 10 minutes to read; weeks to complete!
Activity: Complete the Physical Preparation Checklist

To get ready for a challenge like a big exam, you have to take control of your physical, as well as your mental state. Exercise, proper diet, and rest will ensure that your body works with, rather than against, your mind on test day, as well as during your preparation.

Exercise

If you don't already have a regular exercise program going, the time during which you are preparing for an exam is actually an excellent time to start one. If you are already keeping fit—or trying to get that way—don't let the pressure of preparing for an exam fool you into quitting now. Exercise helps reduce stress by pumping wonderful good-feeling hormones called *endorphins* into your system. It also increases the oxygen supply throughout your body and your brain, so you will be at peak performance on test day.

A half hour of vigorous activity—enough to raise a sweat—every day should be your aim. If you are really pressed for time, every other day is OK. Choose an activity you like and get out there and do it. Jogging with a friend always makes the time go faster as does listening to music.

But don't overdo. You don't want to exhaust yourself. Moderation is the key.

Diet

First of all, cut out the junk. Go easy on caffeine and nicotine, and eliminate alcohol and any other drugs from your system at least two weeks before the exam. Promise yourself a binge the night after the exam, if need be.

What your body needs for peak performance is simply a balanced diet. Eat plenty of fruits and vegetables, along with protein and carbohydrates. Foods that are high in lecithin (an amino acid), such as fish and beans, are especially good "brain foods."

Rest

You probably know how much sleep you need every night to be at your best, even if you don't always get it. Make sure you do get that much sleep, though, for at least a week before the exam. Moderation is important here, too. Extra sleep will just make you groggy.

If you are not a morning person and your exam will be given in the morning, you should reset your internal clock so that your body doesn't think you're taking an exam at 3 A.M. You have to start this process well before the exam. The way it works is to get up half an hour earlier each morning, and then go to bed half an hour earlier that night. Don't try it the other way around; you will just toss and turn if you go to bed early without getting up early. The next morning, get up another half an hour earlier, and so on. How long you will have to do this depends on how late you are used to getting up. Use the *Physical Preparation Checklist* that follows to make sure you are in tip-top form.

▶ STEP 8: GET YOUR ACT TOGETHER

Time to complete: 10 minutes to read; time to complete will vary
Activity: Complete Final Preparations worksheet

Once you feel in control of your mind and body, you are in charge of test anxiety, test preparation, and test-taking strategies. Now, it's time to make charts and gather the materials you need to take to the exam.

Gather Your Materials

The night before the exam, lay out the clothes you will wear and the materials you have to bring with you to the exam. Plan on dressing in layers because you won't have any control over the temperature of the exam room. Have a sweater or jacket you can take off if it's warm. Use the checklist on the worksheet entitled *Final Preparations* on page 29 to help you pull together what you will need.

Don't Skip Breakfast

Even if you don't usually eat breakfast, do so on exam morning. A cup of coffee doesn't count. Don't eat doughnuts or other sweet foods, either. A sugar high will leave you with a sugar low in the middle of the exam. A mix of protein and carbohydrates is best: cereal with milk and just a little sugar or eggs with toast will do your body a world of good.

PHYSICAL PREPARATION CHECKLIST

For the week before the test, write down what physical exercise you engaged in and for how long. Then write down what you ate for each meal. Remember, you are trying for at least half an hour of exercise every other day (preferably every day) and a balanced diet that's light on junk food.

Exam minus 7 days

Exercise: ______ for ______ minutes

Breakfast: ______________________________

Lunch: ______________________________

Dinner: ______________________________

Snacks: ______________________________

Exam minus 6 days

Exercise: ______ for minutes

Breakfast: ______________________________

Lunch: ______________________________

Dinner: ______________________________

Snacks: ______________________________

Exam minus 5 days

Exercise: ______ for ______ minutes

Breakfast: ______________________________

Lunch: ______________________________

Dinner: ______________________________

Snacks: ______________________________

Exam minus 4 days

Exercise: ______ for ______ minutes

Breakfast: ______________________________

Lunch: ______________________________

Dinner: ______________________________

Snacks: ______________________________

Exam minus 3 days
Exercise: ______ for ______ minutes
Breakfast: ____
Lunch: ____
Dinner: ____
Snacks: ____
Exam minus 2 days
Exercise: ______ for ______ minutes
Breakfast: ____
Lunch: ____
Dinner: ____
Snacks: ____
Exam minus 1 day
Exercise: ______ for ______ minutes
Breakfast: ____
Lunch: ____
Dinner: ____
Snacks: ____

STEP 9: DO IT!

Time to complete: 10 minutes, plus test-taking time
Activity: Ace Your Test!

Fast forward to exam day. You are ready. You made a study plan and followed through. You practiced your test-taking strategies while working through this book. You are in control of your physical, mental, and emotional state. You know when and where to show up and what to bring with you. In other words, you are better prepared than most of the other people taking the test with you. You are psyched!

Just one more thing. When you are done with the exam, you will have earned a reward. Plan a celebration. Call your friends and plan a party, or have a nice dinner for two—whatever your heart desires. Give yourself something to look forward to.

And then do it. Go into the exam, full of confidence, armed with test-taking strategies you have practiced until they're second nature. You are in control of yourself, your environment, and your performance on exam day. You are ready to succeed. So do it. Go in there and ace the exam! And, then, look forward to your new career.

FINAL PREPARATIONS

Getting to the Exam Site

Location of exam: ______

Date of exam: ______

Time of exam: ______

Do I know how to get to the exam site? Yes ______ No ______

If no, make a trial run.

Time it will take to get to the exam site: ______

Things to lay out the night before

Clothes I will wear ______

Sweater\jacket ______

Watch ______

Photo ID ______

Admission card ______

4 No. 2 pencils ______

____

____

Arithmetic, Powers, and Roots

ARITHMETIC

Arithmetic is the term used to encompass the following four familiar operations:

- Addition
- Subtraction
- Multiplication
- Division

When solving arithmetic problems, it is helpful to keep in mind the following definitions regarding the operations mentioned above:

- A **sum** is obtained by adding.
- A **difference** is obtained by subtracting.
- A **product** is obtained by multiplying.
- A **quotient** is obtained by dividing.

Basic arithmetic problems require you to add, subtract, multiply, or divide. You may be asked to find the sum, difference, product, or quotient. More advanced arithmetic questions deal with **combined operations**. This simply means that two or more of the basic operations are combined into an equation or expression. For example, a question that asks you to find the product of two sums would be considered a combined operations question.

When dealing with basic arithmetic and combined operations, it is helpful to understand three basic number laws: The commutative law, the associative law, and the distributive law. Sometimes these three laws are referred to as properties (such as the Commutative Property).

- **The commutative law** applies to addition and multiplication and can be represented as $a + b = b + a$ or $a \times b = b \times a$. For example, $2 + 3 = 3 + 2$ and $4 \times 2 = 2 \times 4$ exhibit the commutative law.
- **The associative law** applies to the grouping of addition or multiplication equations and expressions. It can be represented as $a + (b + c) = (a + b) + c$ or $a \times (b \times c) = (a \times b) \times c$. For example, $10 + (12 + 14) = (10 + 12) + 14$.
- **The distributive law** applies to multiplication *over* addition and can be represented as $a(b + c) = ab + ac$. For example, $3(5 + 7) = 3 \times 5 + 3 \times 7$.

It is also especially important to understand the **order of operations**. When dealing with a combination of operations, you must perform the operations in a particular order. An easy way to remember the order of operations is to use the mnemonic **PEMDAS**, where each letter stands for an operation:

- **P**arentheses: Always calculate the values inside the parentheses first.
- **E**xponents: Exponents (or powers) are calculated second.
- **M**ultiplication/**D**ivision: Third, perform any multiplications or divisions in order from left to right.
- **A**ddition/**S**ubtraction: Last, perform any additions or subtractions in order from left to right.

Sample Question:

Two stores are selling the same air conditioner at $357 and $250, respectively. What is the difference in price?

a. $607
b. $170
c. $150
d. $107

The term *difference* means that you will subtract: $357 – $250 = $107. To check your work, just add: $107 + 250 = 357$. The correct answer is **d.**

POWERS

When you raise a number (the base) to an exponent, this is sometimes called raising the number to a *power*.

$\text{Base}^{\text{power}}$ or $\text{Base}^{\text{exponent}}$

When you have the same base, it is easy to combine the exponents according to the following rules:

- When multiplying, such as $a^x \times a^y$, simply add the exponents: $a^x \times a^y = a^{x+y}$
- When dividing, such as $a^x \div a^y$, simply subtract the exponents: $a^x \div a^y = a^{x-y}$
- When raising a power to a power, such as $(a^x)^y$, simply multiply the exponents: $(a^x)^y = a^{x-y}$

Note that if more than one base is included in the parentheses, you must raise all of the bases to the power outside the parentheses, so $(a^x b^y)^z = a^{xz} b^{yz}$.

Two common powers have special names. When raising a number to the 2nd power, it is called squaring the number. When raising a number to the 3rd power, it is called *cubing* the number.

Sample Question:

$(6^2)^5 =$

a. 6^7
b. 6^8
c. 6^{10}
d. 6^{12}

When raising a power to a power, you can just multiply the exponents. Here, you should multiply 2×5, so $(6^2)^5 = 6^{2 \times 5} = 6^{10}$. You can check your work by writing out the solution: $(6^2)^5 = (6 \times 6)^5 = (6 \times 6)(6 \times 6)(6 \times 6)(6 \times 6)(6 \times 6)$. This is 6 to the tenth power. Thus, the correct answer is **c**, 6^{10}.

ROOTS

Typically, you will take the square root of a number. This is denoted by a radical sign, which looks like this: $\sqrt{}$. In order to find the square root of a number, try to figure out what number when squared will equal the number under the radical sign. For example, you know that $2^2 = 4$, so $\sqrt{4} = 2$. Square roots are easy to calculate for *perfect squares*. For example, $\sqrt{4} = 2$, $\sqrt{9} = 3$, $\sqrt{16} = 4$, $\sqrt{25} = 5$, and so forth. Other times, you can approximate the value of a radical by pinpointing it between two perfect squares. For example, since $\sqrt{4} = 2$ and $\sqrt{9} = 3$, $\sqrt{7}$ must be a number between 2 and 3.

In other cases, it is helpful to find equivalents of the radical at hand by applying the rules governing the manipulation of radicals. These rules can be summarized as:

$$\sqrt{ab} = \sqrt{a} \times \sqrt{b}$$

This rule is helpful when simplifying $\sqrt{12}$ for example. $\sqrt{12} = \sqrt{4} \times \sqrt{3} = 2\sqrt{3}$

$$\sqrt{\tfrac{a}{b}} = \sqrt{a} \div \sqrt{b}$$

For example, this rule is helpful when finding the equivalent of $\sqrt{\frac{1}{25}}$. First, take the radical of the top and bottom: $\sqrt{\frac{1}{25}} = \sqrt{1} \div \sqrt{25}$. Since $\sqrt{1} = 1$ and $\sqrt{25} = 5$, we have $\sqrt{1} \div \sqrt{25} = 1 \div 5$.

Once you are able to convert the radicals at hand into equivalents that have the same number under the radical, you can combine them effectively through addition and subtraction. For example, $2\sqrt{2} + 3\sqrt{2} = 5\sqrt{2}$ and $5\sqrt{3} - 4\sqrt{3} = 1\sqrt{3}$.

Sample Question:

$\sqrt{98}$ is equivalent to which of the following?

a. $\sqrt{9} \times \sqrt{8}$
b. $7\sqrt{3}$
c. $49\sqrt{2}$
d. $7\sqrt{2}$

First, look under the radical at 98. Express 98 as 2 factors, trying to make one of the factors a perfect square: $\sqrt{98} = \sqrt{49 \times 2}$. Sometimes it takes a while to get used to figuring out how to rearrange the numbers under the radical. Just remember that if you can find a perfect square, you will be able to pull something out from under the radical. Here, 49 is a perfect square, so we can pull a 7 out from under the radical as follows:

$$\sqrt{49 \times 2} = \sqrt{49} \times \sqrt{2} = 7\sqrt{2}$$

Thus, choice **d** is the correct answer.

PRACTICE QUESTIONS

1. Find the sum of 7,805 and 987.

a. 17,675
b. 8,972
c. 8,987
d. 8,792

2. Lawrence gave $281 to Joel. If he originally had $1,375, how much money does he have left?

a. $1,656
b. $1,294
c. $1,094
d. $984

3. Peter had $10,573 in his savings account. He then deposited $2,900 and $317. How much is in the account now?

a. $13,156
b. $13,790
c. $7,356
d. $6,006

4. What is the positive difference between 10,752 and 675?

a. 11,427
b. 10,077
c. 3,822
d. -10,077

5. 287,500 – 52,988 + 6,808 =

a. 347,396
b. 46,467
c. 333,680
d. 241,320

6. What is the product of 450 and 122?

a. 54,900
b. 6,588
c. 572
d. 328

7. Find the quotient of 12,440 and 40.

a. 497,600
b. 12,480
c. 12,400
d. 311

8. What is the product of 523 and 13 when rounded to the nearest hundred?

a. 6,799
b. 536
c. 6,800
d. 500

9. When the sum of 1,352 and 731 is subtracted from 5,000, the result is
  a. 7,083
  b. 2,917
  c. 2,083
  d. 4,379

10. What is the quotient of 90 divided by 18?
  a. 5
  b. 6
  c. 72
  d. 1,620

11. What is the product of 52 and 22?
  a. 30
  b. 74
  c. 104
  d. 1,144

12. What is the sum of the product of 3 and 2 and the product of 4 and 5?
  a. 14
  b. 26
  c. 45
  d. 90

13. Find the difference of 582 and 73.
  a. 42,486
  b. 655
  c. 509
  d. 408

14. How much greater is the sum of 523 and 65 than the product of 25 and 18?
  a. 138
  b. 545
  c. 588
  d. 33,545

15. Solve 589 + 7,995 ÷ 15.
  a. 572 with a remainder of 4
  b. 1,122
  c. 8,569
  d. 8,599

16. $540 \div 6 + 3 \times 24 =$
- **a.** 2,232
- **b.** 1,440
- **c.** 1,260
- **d.** 162

17. $78 \times (32 + 12) =$
- **a.** 2,508
- **b.** 3,432
- **c.** 6,852
- **d.** 29,953

18. Which of the following demonstrates the commutative property?
- **a.** $2 + 3 = 4 + 1$
- **b.** $2 + (3 + 4) = (2 + 3) + 4$
- **c.** $2 \times 3 = 3 \times 2$
- **d.** $2 \times (3 \times 4) = (2 \times 3) \times 4$

19. Which of the following demonstrates the associative property?
- **a.** $4 + 5 = 5 + 4$
- **b.** $2 \times (3 + 4) = (2 \times 3) + 4$
- **c.** $4 \times 5 = 5 \times 4$
- **d.** $2 \times (3 \times 4) = (2 \times 3) \times 4$

20. Which of the following demonstrates the distributive property?
- **a.** $(4 \times 5) + 1 = 4 \times (5 + 1)$
- **b.** $4 \times (5 + 1) = 4 \times 5 + 4 \times 1$
- **c.** $4 \times 5 \times 1 = 1 \times 5 \times 4$
- **d.** $(4 + 5) + 1 = 4 + (5 + 1)$

21. $4 \times 4 \times 4 \times 4$ is equivalent to
- **a.** 4×4^2
- **b.** $4^2 \times 4^3$
- **c.** $(4^2)^2$
- **d.** $4^3 + 4^2$

22. What is the square root of 81?
- **a.** 8
- **b.** 9
- **c.** 10
- **d.** 11

23. $11^3 =$

a. 121
b. 1,331
c. 14,641
d. 15,551

24. $(8^3)^5$ is equal to

a. 8^{15}
b. 8^8
c. 8^4
d. 8^2

25. $\sqrt{72}$ is equivalent to

a. 12
b. $6\sqrt{3}$
c. $6\sqrt{2}$
d. $36\sqrt{2}$

26. 7^3 is equal to

a. 343
b. 49
c. 38
d. 21

27. $2\sqrt{128}$ is equivalent to

a. $8\sqrt{2}$
b. $16\sqrt{2}$
c. $32\sqrt{2}$
d. $64\sqrt{2}$

28. $\sqrt{50} + \sqrt{162} =$

a. $106\sqrt{2}$
b. $14\sqrt{2}$
c. $9\sqrt{2}$
d. $5\sqrt{2}$

29. $75 - 3\ (9-7)^4 =$

a. 3^3
b. 144^4
c. 69^4
d. 5^4

30. $\sqrt{1{,}225} =$

a. 30
b. 35
c. 40
d. 45

31. $3 \times 3 \times 3 \times 3 \times 3 \times 3$ is equivalent to

a. $(3^3)^3$
b. $3^2 \times 3^2 \times 3^2$
c. $3^2 \times 3^3$
d. $(3^4)^2$

32. $2\sqrt{3} + 2\sqrt{2} + 5\sqrt{3} =$

a. $4\sqrt{3} + 2\sqrt{2}$
b. $4\sqrt{2} + 5\sqrt{3}$
c. $8\sqrt{2} + 2\sqrt{3}$
d. $7\sqrt{3} + 2\sqrt{2}$

33. $\sqrt{\frac{1}{81}} =$

a. $1 \div 9$
b. $1 \div 81$
c. $1 \div \sqrt{3}$
d. $1 \div \sqrt{9}$

34. $(-3)^3 + (3)^3$ is equivalent to

a. 54
b. 27
c. 0
d. −27

35. $\sqrt{70}$ is between which of the following two numbers?

a. 5 and 6
b. 6 and 7
c. 7 and 8
d. 8 and 9

36. 18^3 is how much greater than 16^2?

a. 6,088
b. 5,576
c. 265
d. 68

37. 42^2 is how much greater than 24^2?
- **a.** 1,188
- **b.** 1,764
- **c.** 576
- **d.** 2,340

38. $\sqrt{(-3)^2(4)^2}$ is equivalent to
- **a.** $12\sqrt{2}$
- **b.** $-\sqrt{12^2}$
- **c.** 12
- **d.** –12

39. $(-12)^2 =$
- **a.** –144
- **b.** –121
- **c.** 121
- **d.** 144

40. $(-3)^3 =$
- **a.** 9
- **b.** –9
- **c.** 27
- **d.** –27

41. The square root of 48 is between which two numbers?
- **a.** 6 and 7
- **b.** 5 and 6
- **c.** 4 and 5
- **d.** 3 and 4

42. $2^4 \times 2^7$ is equivalent to
- **a.** 2^{28}
- **b.** 2^{11}
- **c.** 2^5
- **d.** 2^3

43. $3^2 + 3^3 =$
- **a.** 18
- **b.** 27
- **c.** 6^2
- **d.** 6^5

44. $7^{11} \div 7^9 =$

a. 7^{20}
b. 7^{-20}
c. 49
d. $1 \div 49$

45. $3^5 \times 3^2 \times 5^3 \times 5^9 =$

a. $3^7 \times 5^{12}$
b. $3^{12} \times 5^7$
c. $3^3 \times 5^6$
d. $3^6 \times 5^3$

46. $(6^9 \times 2^5) \div (6^8 \times 2^2)$ is equivalent to

a. 64
b. 48
c. 32
d. 16

47. Solve: $\frac{10 \times 10^{10}}{5 \times 10^2} =$

a. 10×10^8
b. 5×10^{-8}
c. 2×10^8
d. 5×10^8

48. Find the sum of 3×10^2 and 2×10^5.

a. 200,300
b. 23,000
c. 2,300
d. 230

49. What is the product of 2×10^6 and 6×10^7?

a. 12×10^{42}
b. 12×10^{13}
c. 12×10^5
d. 12×10^3

50. A rod that is 8×10^6 mm is how much longer than a rod that is 4×10^4 mm?

a. twice as large
b. four times as large
c. twenty times as large
d. two hundred times as large

ANSWERS

1. **d.** *Sum* means addition, so 7,805 + 987 = 8,792. The correct answer is **d.**
2. **c.** To find the difference, just subtract: 1,375 – 281 = 1,094. He now has $1,094.
3. **b.** Add all three values together: 10,573 + 2,900 + 317 = $13,790.
4. **b.** To find a *difference*, just subtract. The term *positive difference* means you are solving for a positive answer. This means you should subtract the smaller number from the larger number: 10,752 – 675 = 10,077.
5. **d.** 287,500 – 52,988 = 234,512. Next, add: 234,512 + 6,808 = 241,320.
6. **a.** *Product* means multiply. 450 × 122 = 54,900.
7. **d.** A quotient results from division. 12,440 ÷ 40 = 311.
8. **c.** To find the product, just multiply: 523 × 13 = 6,799. Rounding to the nearest hundred yields 6,800.
9. **b.** The sum of 1,352 and 731 is obtained by adding: 1,352 + 731 = 2,083. Next, we subtract this value from 5,000: 5,000 – 2,083 = 2,917.
10. **a.** 90 divided by 18 = 5. Thus, the quotient is 5.
11. **d.** The product is obtained by multiplying: 52 × 22= 1,144.
12. **b.** First, find the 2 products:
    3 × 2 = 6 and 4 × 5 = 20.
    Next, add these 2 products together: 6 + 20 = 26.
13. **c.** To find a difference, you subtract: 582 – 73 = 509.
14. **a.** First, calculate the two equations:
    The sum of 523 and 65: 523 + 65 = 588
    The product of 25 and 18: 25 × 18 = 450
    Next, find the difference:
    588 – 450 = 138
15. **b.** The rules for the order of operations state that division should be done before addition. Recall **PEMDAS:** *parentheses, exponents, multiplication, division, addition, subtraction.* 7,995 ÷ 15 = 533. Next add 589 + 533 = 1,122.
16. **d.** Consider **PEMDAS**: *parentheses, exponents, multiplication, division, addition, subtraction.* Here, you must solve the division first: 540 ÷ 6 = 90. The equation becomes 90 + 3 × 24. Again, considering PEMDAS you know you should calculate the multiplication first. 3 × 24 = 72, so the equation reduces to 90 + 72 = 162.
17. **b.** Consider **PEMDAS:** *parentheses, exponents, multiplication, division, addition, subtraction.* Here, you must solve the part inside the parentheses first: 32 + 12 = 44. The equation becomes 78 × 44. Multiplying, you get: 3,432.
18. **c.** Note that this question is not looking for a true equation. It is asking which equation represents the commutative property. The commutative property applies for addition and multiplication and can be represented as $a + b = b + a$ or $a \times b = b \times a$. Choice **c** shows this relationship: 2 × 3 = 3 × 2. In other words, the order in which you multiply two numbers does not matter.

19. **d.** The associative property applies to grouping of addition or multiplication problems. It can be represented as $a + (b + c) = (a + b) + c$ or $a \times (b \times c) = (a \times b) \times c$. Note that you CANNOT combine addition and multiplication as in choice **b.** $2 \times (3 + 4) \neq (2 \times 3) + 4$. Only choice **d** correctly shows this property: $2 \times (3 \times 4) = (2 \times 3) \times 4$.

20. **b.** The distributive property applies to multiplication over addition such as in choice **b**: $4 \times (5 + 1) = 4 \times 5 + 4 \times 1$. Notice that multiplying the sum of the two terms by 4 is equivalent to multiplying each term by 4 and then adding these values.

21. **c.** $4 \times 4 \times 4 \times 4$ is the same as 4^4. Choice **c** also equals 4^4 because when you raise a power to another power you simply multiply the exponents. Thus, $(4^2)^2 = 4^{2\times2}$. Choice **a** equals 4^3, choice **b** equals 4^5, and choice **d** equals 64 + 16, or 80.

22. **b.** The square root of 81 simply means $\sqrt{81}$. To solve, just ask yourself, "What number squared equals 81?" $9^2 = 81$, so $\sqrt{81} = 9$.

23. **b.** $11^3 = 11 \times 11 \times 11 = 121 \times 11 = 1{,}331$.

24. **a.** When raising a power of a base to another power, you just multiply the exponents. Here $(8^3)^5 = 8^{3\times5} = 8^{15}$.

25. **c.** $\sqrt{72} = \sqrt{36 \times 2}$. Because $36 = 6^2$, you can pull a 6 out from under the radical. Thus, you have, $6\sqrt{2}$.

26. **a.** $7^3 = 7 \times 7 \times 7$ which equals $49 \times 7 = 343$.

27. **b.** $2\sqrt{128}$ is equal to $2\sqrt{64 \times 2}$, or $2 \times \sqrt{64} \times \sqrt{2}$. Since $\sqrt{64} = 8$, we have $2 \times 8 \times \sqrt{2} = 16\sqrt{2}$.

28. **b.** Each radical can be rewritten. First, $\sqrt{50} = \sqrt{2} \times \sqrt{25} = \sqrt{2} \times \sqrt{25} = \sqrt{2} \times 5 = 5\sqrt{2}$. Next, $\sqrt{162} = \sqrt{81 \times 2} = \sqrt{81} \times \sqrt{2} = 9\sqrt{2}$. Finally, add the 2 radicals: $5\sqrt{2} + 9\sqrt{2} = 14\sqrt{2}$.

29. **a.** Consider PEMDAS: *parentheses, exponents, multiplication, division, addition, subtraction.* First, calculate the value inside the parentheses: $75 - 3\,(9-7)^4 = 75 - 3\,(2)^4$. Second, calculate the exponent $75 - 3\,(2)^4 = 75 - 3\,(16)$. Third, calculate the multiplication: $75 - 3(16) = 75 - 48$. Finally, subtract: $75 - 48 = 27$. Because 27 is not listed as an answer choice, figure out which choice equals 27. Here, choice **a**, $3^3 = 3 \times 3 \times 3 = 27$.

30. **b.** In this case, it is easiest to see which answer choice when squared equals 1,225. Choice **a**, 30, would yield $30 \times 30 = 900$, and is thus too small. Choice **b**, 35 yields $35 \times 35 = 1{,}225$. Thus $\sqrt{1{,}225} = 35$ and choice **b** is correct.

31. **b.** $3 \times 3 \times 3 \times 3 \times 3 \times 3$ is equivalent to 3^6. Choice **b** is equivalent to 3^6 because $3^2 \times 3^2 \times 3^2$ equals 3^{2+2+2}. Remember to add the powers when multiplying numbers with the same base. Choice **a** equals 3^9, choice **c** equals 3^5, and choice **d** equals 3^8.

32. **d.** You can combine the two terms with the $\sqrt{3}$. $2\sqrt{3} + 5\sqrt{3} = 7\sqrt{3}$, so the entire expression equals $7\sqrt{3} + 2\sqrt{2}$.

33. **a.** $\sqrt{\frac{1}{81}} = \sqrt{1} \div \sqrt{81} = 1 \div 9$, choice **a**.

34. **c.** Cubing a negative number (or taking any odd power of a negative number for that matter) results in a negative value. Here, $^-3^3 = {^-3} \times {^-3} \times {^-3} = {^-27}$. $3^3 = 27$. Thus, the sum $(-3)^3 + (3)^3 = -27 + 27 = 0$.

35. **d.** 8^2 is 64 and 9^2 is 81. Thus, the square root of 70 (which is between 64 and 81) must be between 8 and 9.

36. **b.** First, calculate both quantities: $18^3 = 18 \times 18 \times 18 = 5{,}832$ and $16^2 = 256$. Next, in order to find out how much greater the first quantity is, we find the *difference* (by subtracting): $5{,}832 - 256 = 5{,}576$.

37. **a.** Calculate both of the given quantities: $42^2 = 1{,}764$ and $24^2 = 576$. Next, subtract to obtain the difference: $1{,}764 - 576 = 1{,}188$.

38. **c.** To solve $\sqrt{(-3)^2(4)^2}$ we will first simplify the value under the radical. $(-3)^2 = 9$ and $4^2 = 16$, so $\sqrt{(-3)^2(4)^2} = \sqrt{9 \times 16}$. This can be rewritten as $\sqrt{9} \times \sqrt{16}$ and simplified to 3×4, which equals 12.

39. **d.** When you square a negative number (or raise a negative number to any even power) the result is a positive number. So, $(-12)^2 = 144$.

40. **d.** When you raise a negative number to any odd power, the result is a negative number. So, $(-3)^3 = -3 \times -3 \times -3 = -27$.

41. **a.** $6^2 = 36$ and $7^2 = 49$. So, radical 48 (which is between 36 and 49) will equal a number that is between 6 and 7.

42. **b.** Since the base (2) is the same, you can simply add the exponents. $2^4 \times 2^7 = 2^{4+7} = 2^{11}$.

43. **c.** $3^2 = 9$ and $3^3 = 27$. $9 + 27 = 36$. Because 36 is not listed as an answer choice, calculate which choice equals 36. Here, choice **c**, $6^2 = 6 \times 6 = 36$, and is thus correct.

44. **c.** Since the base (7) is the same, you can simply subtract the exponents. $7^{11} \div 7^9 = 7^{11-9} = 7^2 = 49$.

45. **a.** You can apply the rules of exponents to the terms that have the same bases. Thus, $3^5 \times 3^2 \times 5^3 \times 5^9 = 3^{5+2} \times 5^{3+9} = 3^7 \times 5^{12}$.

46. **b.** You can apply the rules of exponents to the terms that have the same bases. Thus, $(6^9 \times 2^5) \div (6^8 \times 2^2)$ is equivalent to $6^{9-8} \times 2^{5-2} = 6^1 \times 2^3 = 6 \times 8 = 48$.

47. **c.** $\frac{10 \times 10^{10}}{5 \times 10^2} = \frac{10}{5} \times \frac{10^{10}}{10^2} = 2 \times 10^{10-2} = 2 \times 10^8$

Remember, according to the rules of exponents, when dividing, you can simply subtract the exponents of the 2 powers of 10.

48. **a.** $3 \times 10^2 = 3 \times 100 = 300$ and $2 \times 10^5 = 2 \times 100{,}000 = 200{,}000$. Adding these 2 values yields $200{,}000 + 300 = 200{,}300$.

49. **b.** The product of 2×10^6 and 6×10^7 would be $2 \times 10^6 \times 6 \times 10^7 = 2 \times 6 \times 10^6 \times 10^7$. Applying the rules of exponents, you can simply add the exponents of the 2 powers of 10. Thus, $2 \times 6 \times 10^6 \times 10^7 = 2 \times 6 \times 10^{6+7} = 2 \times 6 \times 10^{13}$. Multiplying the first 2 terms yields 12×10^{13}.

50. **d.** 8×10^6 mm $= 8 \times 1{,}000{,}000 = 8{,}000{,}000$ mm. 4×10^4 mm $= 4 \times 10{,}000 = 40{,}000$. How many times larger is 8,000,000 than 40,000? $8{,}000{,}000 \times 40{,}000 = 200$. Thus, the first rod is 200 times larger than the second.

Fractions

Problems involving fractions may be straightforward calculation questions, or they may be word problems. Typically, they ask you to add, subtract, multiply, divide, or compare fractions.

WORKING WITH FRACTIONS

A fraction is a part of a whole. Fractions are written as part/whole, or more technically as numerator/denominator.

THREE KINDS OF FRACTIONS

Proper fraction: The top number is less than the bottom number:

$\frac{1}{2}$; $\frac{2}{3}$; $\frac{4}{9}$; $\frac{8}{13}$

The value of a proper fraction is less than 1.

Improper fraction: The top number is greater than or equal to the bottom number:

$\frac{3}{2}$; $\frac{5}{3}$; $\frac{14}{9}$; $\frac{12}{12}$

The value of an improper fraction is 1 or more.

Mixed number: A fraction written to the right of a whole number:

$3\frac{1}{2}$; $4\frac{2}{3}$; $7\frac{2}{3}$; $12\frac{3}{4}$; $24\frac{3}{4}$

The value of a mixed number is more than 1: it is the sum of the whole number plus the fraction.

CHANGING IMPROPER FRACTIONS INTO MIXED OR WHOLE NUMBERS

It's easier to add and subtract fractions that are mixed numbers rather than improper fractions. To change an improper fraction, say $\frac{13}{2}$, into a mixed number, follow these steps:

1. Divide the bottom number (2) into the top number (13) to get the whole number portion (6) of the mixed number:

$$\begin{array}{r} 6 \\ 2\overline{)13} \\ -12 \\ \hline 1 \end{array}$$

2. Write the remainder of the division (1) over the old bottom number (2): $6\frac{1}{2}$
3. Check: Change the mixed number back into an improper fraction (see steps below).

CHANGING MIXED NUMBERS INTO IMPROPER FRACTIONS

It's easier to multiply and divide fractions when you are working with improper fractions rather than mixed numbers. To change a mixed number, say $2\frac{3}{4}$, into an improper fraction, follow these steps:

1. Multiply the whole number (2) by the bottom number (4): $2 \times 4 = 8$
2. Add the result (8) to the top number (3): $8 + 3 = 11$
3. Put the total (11) over the bottom number (4): $\frac{11}{4}$
4. Check: Reverse the process by changing the improper fraction into a mixed number. If you get back the number you started with, your answer is right.

REDUCING FRACTIONS

Reducing a fraction means writing it in lowest terms, that is, with smaller numbers. For instance, 50¢ is $\frac{50}{100}$ of a dollar, or $\frac{1}{2}$ of a dollar. In fact, if you have 50¢ in your pocket, you say that you have half a dollar. Reducing a fraction does not change its value.

Follow these steps to reduce a fraction:

1. Find a whole number that divides evenly into both numbers that make up the fraction.
2. Divide that number into the top of the fraction, and replace the top of the fraction with the quotient (the answer you got when you divided).
3. Do the same thing to the bottom number.
4. Repeat the first three steps until you can't find a number that divides evenly into both numbers of the fraction.

For example, let's reduce $\frac{8}{24}$. We could do it in 2 steps: $\frac{8 \div 4}{24 \div 4} = \frac{2}{6}$; then $\frac{2 \div 2}{6 \div 2} = \frac{1}{3}$.
Or we could do it in a single step: $\frac{8 \div 8}{24 \div 8} = \frac{1}{3}$.

Shortcut: When the top and bottom numbers both end in zeros, cross out the same number of zeros in both numbers to begin the reducing process. For example, $\frac{300}{4{,}000}$ reduces to $\frac{3}{40}$ when you cross out two zeros in both numbers.

Whenever you do arithmetic with fractions, reduce your answer. On a multiple-choice test, don't panic if your answer isn't listed. Try to reduce it and then compare it to the choices.

Sample Question:

The fraction $\frac{80}{200}$ is equivalent to which of the following?

a. $\frac{4}{100}$
b. $\frac{2}{5}$
c. $\frac{3}{4}$
d. $\frac{1}{4}$

You can reduce this fraction in steps: $\frac{80}{200} \div \frac{20}{20} = \frac{4}{10} \div \frac{2}{2} = \frac{2}{5}$. Thus, choice **b** is correct.

▶ RAISING FRACTIONS TO HIGHER TERMS

Before you can add and subtract fractions, you have to know how to raise a fraction to higher terms. This is actually the opposite of reducing a fraction.

Follow these steps to raise $\frac{2}{3}$ to 24ths:

1. Divide the old bottom number (3) into the new one (24): $24 \div 3 = 8$
2. Multiply the answer (8) by the old top number (2): $2 \times 8 = 16$
3. Put the answer (16) over the new bottom number (24): $\frac{16}{24}$
4. Check: Reduce the new fraction to see if you get the original number back: $\frac{16 \div 8}{24 \div 8} = \frac{2}{3}$

ADDING FRACTIONS

If the fractions have the same bottom numbers, just add the top numbers together and write the total over the bottom number.

Example: $\frac{2}{9} + \frac{4}{9} = \frac{2+4}{9} = \frac{6}{9}$
Reduce the sum: $\frac{2}{3}$

Example: $\frac{5}{8} + \frac{7}{8} = \frac{12}{8}$
Change the sum to a mixed number: $1\frac{4}{8}$; then reduce: $1\frac{1}{2}$

There are a few extra steps to add mixed numbers with the same bottom numbers, say $2\frac{3}{5} + 1\frac{4}{5}$:

1. Add the fractions: $\frac{3}{5} + \frac{4}{5} = \frac{7}{5}$
2. Change the improper fraction into a mixed number: $\frac{7}{5} = 1\frac{2}{5}$
3. Add the whole numbers: $2 + 1 = 3$
4. Add the results of steps 2 and 3: $1\frac{2}{5} + 3 = 4\frac{2}{5}$

FINDING A COMMON DENOMINATOR

If the fractions you want to add don't have the same bottom number, you will have to raise some or all of the fractions to higher terms so that they all have the same bottom number, the *common denominator*.

See if all the bottom numbers divide evenly into the biggest bottom number. Check out the multiplication table of the largest bottom number until you find a number that all the other bottom numbers evenly divide into. When all else fails, multiply all the bottom numbers together.

Example: $\frac{2}{3} + \frac{4}{5}$

1. Find the common denominator. Multiply the bottom numbers: $3 \times 5 = 15$
2. Raise each fraction to 15ths: $\frac{2}{3} = \frac{10}{15}$
   $\frac{4}{5} = \frac{12}{15}$
3. Add as usual: $\frac{22}{15}$

FINDING THE LEAST COMMON DENOMINATOR

If you are asked to find the least common denominator (the LCD), you will need to find the smallest number that is a multiple of the original denominators present. Sometimes you can figure this out mentally, or you will stumble onto the LCD by following the steps above.

However, to be sure that you have the *least* common denominator, you can use one of two methods:

1. Find the least common multiple. This can be done by checking out the multiplication table of the largest bottom number until you find a number that all the other bottom numbers evenly divide into, as described above.
2. Determine the prime factorization of each of the denominators. The least common denominator will encompass every denominator's prime factorization.

Prime numbers are numbers that have only two factors, the number 1 and itself. For example, 3 is prime because it's only factors are 1 and 3. Numbers that are not prime can be expressed in terms of prime factors. For example, let's compute the prime factorization of 12.

$12 = 3 \times 4 = 3 \times 2 \times 2$

Thus, the prime factorization of 12 is $3 \times 2 \times 2$.

In order to find the LCD of $\frac{3}{4}$ and $\frac{5}{6}$, we can use the prime factorization method as follows:

1. Find the prime factorization of both denominators:
   $4 = 2 \times 2$
   $6 = 2 \times 3$
2. The LCD will contain the prime factorization of both denominators:
   $4 = 2 \times 2$ the LCD must have two 2s
   $6 = 2 \times 3$ the LCD must have a 2 and a 3

The LCD will be $2 \times 2 \times 3$. Note that this LCD contains the prime factorization of 4 and 6.

Sample Question: $\frac{4}{5} + \frac{1}{6} =$

a. $\frac{5}{6}$

b. $\frac{5}{11}$

c. $\frac{7}{15}$

d. $\frac{29}{30}$

The quickest way to find a common denominator is to multiply the two given denominators. 5 × 6 = 30, so the new denominator will be 30. To convert $\frac{4}{5}$ into 30ths, we multiply by $\frac{6}{6}$: $\frac{4}{5} \times \frac{6}{6} = \frac{24}{30}$. To convert $\frac{1}{6}$ into 30ths, we multiply by $\frac{5}{5}$: $\frac{1}{6} \times \frac{5}{5} = \frac{5}{30}$. Next, we add: $\frac{24}{30} + \frac{5}{30} = \frac{29}{30}$. Thus, the correct answer is **d.**

▶ SUBTRACTING FRACTIONS

If the fractions have the same bottom numbers, just subtract the top numbers and write the difference over the bottom number.

Example: $\frac{4}{9} - \frac{3}{9} = \frac{4-3}{9} = \frac{1}{9}$

If the fractions you want to subtract don't have the same bottom number, you will have to raise some or all of the fractions to higher terms so that they all have the same bottom number, or LCD. If you forgot how to find the LCD, just read the section on adding fractions with different bottom numbers.

Example: $\frac{5}{6} - \frac{3}{4}$

1. Raise each fraction to 12ths because 12 is the LCD, the smallest number $\frac{5}{6} = \frac{10}{12}$ that 6 and 4 both divide into evenly: $\frac{3}{4} = \frac{9}{12}$
2. Subtract as usual: $\frac{1}{12}$

Subtracting mixed numbers with the same bottom number is similar to adding mixed numbers.

Example: $4\frac{3}{5} - 1\frac{2}{5}$

1. Subtract the fractions: $\frac{3}{5} - \frac{2}{5} = \frac{1}{5}$
2. Subtract the whole numbers: 4 – 1 = 3
3. Add the results of steps 1 and 2: $\frac{1}{5} + 3 = 3\frac{1}{5}$

Sometimes there is an extra borrowing step when you subtract mixed numbers with the same bottom numbers, say $7\frac{3}{5} - 2\frac{4}{5}$:

1. You can't subtract the fractions the way they are because $\frac{4}{5}$ is bigger than $\frac{3}{5}$. So you borrow 1 from the 7, making it 6, and change that 1 to $\frac{5}{5}$ because 5 is the bottom number: $7\frac{3}{5} = 6\frac{5}{5} + \frac{3}{5}$
2. Add the numbers from step 1: $6\frac{5}{5} + \frac{3}{5} = 6\frac{8}{5}$
3. Now, you have a different version of the original problem: $6\frac{8}{5} - 2\frac{4}{5}$
4. Subtract the fractional parts of the two mixed numbers: $\frac{8}{5} - \frac{4}{5} = \frac{4}{5}$
5. Subtract the whole number parts of the two mixed numbers: 6 – 2 = 4
6. Add the results of the last two steps together: $4 + \frac{4}{5} = 4\frac{4}{5}$

MULTIPLYING FRACTIONS

Multiplying fractions is actually easier than adding them. All you do is multiply the top numbers and then multiply the bottom numbers.

For example, $\frac{2}{3} \times \frac{5}{7} = \frac{2 \times 5}{3 \times 7} = \frac{10}{21}$

Sometimes you can cancel before multiplying. Cancelling is a shortcut that makes the multiplication go faster because you are multiplying with smaller numbers. It's very similar to reducing: if there is a number that divides evenly into a top number and bottom number, do that division before multiplying. If you forget to cancel, you will still get the right answer, but you will have to reduce it.

Example: $\frac{5}{6} \times \frac{9}{20}$

1. Cancel the 6 and the 9 by dividing 3 into both of them: $6 \div 3 = 2$ and $9 \div 3 = 3$. Cross out the 6 and the 9:

   $\frac{5}{\cancel{6}_2} \times \frac{\cancel{9}^3}{20}$

2. Cancel the 5 and the 20 by dividing 5 into both of them: $5 \div 5 = 1$ and $20 \div 5 = 4$. Cross out the 5 and the 20:

   $\frac{\cancel{5}^1}{2} \times \frac{3}{\cancel{20}_4}$

3. Multiply across the new top numbers and the new bottom numbers: $\frac{1 \times 3}{2 \times 4} = \frac{3}{8}$

To multiply a fraction by a whole number, first rewrite the whole number as a fraction with a bottom number of 1:

Example: $5 \times \frac{2}{3} = \frac{5}{1} \times \frac{2}{3} = \frac{10}{3}$ (Optional: convert $\frac{10}{3}$ to a mixed number: $3\frac{1}{3}$)

To multiply with mixed numbers, it's easier to change them to improper fractions before multiplying.

Example: $4\frac{2}{3} \times 5\frac{1}{2}$

1. Convert $4\frac{2}{3}$ to an improper fraction: $4\frac{2}{3} = \frac{4 \times 3 + 1}{3} = \frac{14}{3}$
2. Convert $5\frac{1}{2}$ to an improper fraction: $5\frac{1}{2} = \frac{5 \times 2 + 1}{2} = \frac{11}{2}$
3. Cancel and multiply the fractions:
4. Optional: convert the improper fraction to a mixed number: $\frac{77}{3} = 25\frac{2}{3}$

Tip: When you find a fraction of a number, you just find the product of the two numbers.

Sample Problem: What is $4\frac{1}{5}$ of $2\frac{1}{3}$?

a. $9\frac{4}{5}$

b. $8\frac{1}{15}$

c. $8\frac{1}{5}$

d. $8\frac{1}{3}$

To find $4\frac{1}{5}$ of $2\frac{1}{3}$, you just find the product (multiply). First, convert both fractions into improper fractions. $4\frac{1}{5} = \frac{4 \times 5 + 1}{5} = \frac{21}{5}$; $2\frac{1}{3} = \frac{2 \times 3 + 1}{3} = \frac{7}{3}$. Next, multiply: $\frac{21}{5} \times \frac{7}{3}$. Note that you can cancel:

$$\frac{\overset{7}{\cancel{21}}}{5} \times \frac{7}{\underset{1}{\cancel{3}}} = \frac{49}{5} = 9\frac{4}{5}$$

Thus, choice **a** is correct.

▶ DIVIDING FRACTIONS

To divide one fraction by a second fraction, invert the second fraction (that is, flip the top and bottom numbers) and then multiply. That's all there is to it!

Example: $\frac{1}{2} \div \frac{3}{5}$

1. Invert the second fraction ($\frac{3}{5}$): $\frac{5}{3}$
2. Change the division sign ($\div$) to a multiplication sign: ($\times$)
3. Multiply the first fraction by the new second fraction: $\frac{1}{2} \times \frac{5}{3} = \frac{1 \times 5}{2 \times 3} = \frac{5}{6}$

To divide a fraction by a whole number, first change the whole number to a fraction by putting it over 1. Then follow the division steps above.

Example: $\frac{3}{5} \div 2 = \frac{3}{5} \div \frac{2}{1} = \frac{3}{5} \times \frac{1}{2} = \frac{3 \times 1}{5 \times 2} = \frac{3}{10}$

When the division problem has a mixed number, convert it to an improper fraction and then divide as usual.

Example: $2\frac{3}{4} \div \frac{1}{6}$

1. Convert $2\frac{3}{4}$ to an improper fraction: $2\frac{3}{4} = \frac{2 \times 4 + 3}{4} = \frac{11}{4}$
2. Divide $\frac{11}{4}$ by $\frac{1}{6}$: $\frac{11}{4} \div \frac{1}{6} = \frac{11}{4} \times \frac{6}{1}$
3. Flip $\frac{1}{6}$ to $\frac{6}{1}$, change $\div$ to $\times$, cancel and multiply:

$$\frac{11}{\underset{2}{\cancel{4}}} \times \frac{\overset{3}{\cancel{6}}}{1} = \frac{33}{2}$$

Sample Problem: $5\frac{1}{6} \div 7\frac{1}{2} =$

a. $\frac{465}{12}$

b. $\frac{1}{3}$

c. $\frac{1}{15}$

d. $\frac{31}{45}$

First, you should convert the mixed numbers into improper fractions. $5\frac{1}{6} = \frac{5 \times 6 + 1}{6}$ and $7\frac{1}{2} = \frac{7 \times 2 + 1}{2} = \frac{15}{2}$. So far you have: $\frac{31}{6} \div \frac{15}{2}$. Next, rewrite this as a multiplication problem: $\frac{31}{6} \times \frac{2}{15} = \frac{62}{90}$. Finally, reduce this fraction: $\frac{62}{90} \div \frac{2}{2} = \frac{31}{45}$. Thus, choice **d** is correct.

PRACTICE QUESTIONS

1. What is the LCD of $\frac{1}{12}$, $\frac{4}{9}$, $\frac{5}{18}$, and $\frac{1}{24}$?

a. 24

b. 48

c. 216

d. 46,656

2. What is the sum of $\frac{2}{9}$ and $\frac{5}{9}$?

a. $\frac{4}{9}$

b. $\frac{7}{9}$

c. $\frac{1}{3}$

d. $\frac{2}{3}$

3. $3\frac{4}{5}$ is equal to which of the following improper fractions?

a. $\frac{12}{5}$

b. $\frac{17}{5}$

c. $\frac{19}{5}$

d. $\frac{23}{5}$

4. $\frac{5}{6} - \frac{1}{6} =$

a. $\frac{4}{5}$

b. $\frac{2}{3}$

c. $\frac{1}{4}$

d. $\frac{1}{8}$

5. Convert $\frac{29}{3}$ into a mixed number.

a. $9\frac{2}{3}$
b. $8\frac{2}{3}$
c. $9\frac{1}{3}$
d. $8\frac{1}{3}$

6. Find the sum of $\frac{4}{9}$ and $\frac{3}{4}$.

a. $\frac{11}{36}$
b. $\frac{42}{36}$
c. $1\frac{7}{36}$
d. $1\frac{1}{6}$

7. Find the sum of $1\frac{2}{5}$ and $\frac{2}{9}$.

a. $1\frac{28}{45}$
b. $3\frac{4}{14}$
c. $1\frac{4}{14}$
d. $1\frac{2}{3}$

8. Reduce the following fraction to its simplest form: $\frac{9}{54}$

a. $\frac{1}{8}$
b. $\frac{1}{6}$
c. $\frac{3}{18}$
d. $\frac{3}{16}$

9. Change $\frac{154}{11}$ to a whole number.

a. 8
b. 14
c. 18
d. 32

10. What is the sum of $15\frac{1}{4}$, $9\frac{2}{3}$, $7\frac{1}{5}$, and $23\frac{1}{2}$?

a. $54\frac{1}{15}$
b. $55\frac{7}{30}$
c. $55\frac{37}{60}$
d. $56\frac{1}{2}$

11. The mixed number $8\frac{2}{3}$ is equivalent to which improper fraction below?

a. $\frac{6}{3}$
b. $\frac{10}{3}$
c. $\frac{16}{3}$
d. $\frac{26}{3}$

12. Express $\frac{20}{4}$ as a whole number.

a. 8
b. 5
c. 4
d. 3

13. Convert $2\frac{5}{8}$ to an improper fraction.

a. $\frac{21}{8}$
b. $\frac{7}{8}$
c. $\frac{25}{8}$
d. $\frac{5}{16}$

14. The reciprocal of $1\frac{1}{3}$ is

a. $\frac{4}{3}$
b. 1
c. $-1\frac{1}{3}$
d. $\frac{3}{4}$

15. Which of the following choices is an improper fraction?

a. $\frac{16}{24}$
b. $\frac{95}{37}$
c. $\frac{23}{90}$
d. $\frac{11}{80}$

16. Subtract $13\frac{1}{5}$ from $22\frac{1}{4}$.

a. $9\frac{1}{4}$
b. $9\frac{2}{5}$
c. $9\frac{1}{15}$
d. $9\frac{1}{20}$

17. Which of the following has the greatest value?

a. $\frac{7}{18}$

b. $\frac{10}{16}$

c. $\frac{5}{12}$

d. $\frac{3}{4}$

18. What is the product of $18\frac{1}{5}$ and 35?

a. 630

b. 637

c. $640\frac{1}{5}$

d. $645\frac{2}{5}$

19. What is the LCD of $\frac{2}{3}$, $\frac{3}{4}$, and $\frac{5}{16}$?

a. 192

b. 64

c. 48

d. 16

20. Express this improper fraction as a mixed number: $\frac{23}{5}$

a. $4\frac{3}{5}$

b. $5\frac{2}{5}$

c. $4\frac{2}{5}$

d. $5\frac{3}{5}$

21. What is the LCD of $\frac{7}{18}$, $\frac{3}{4}$, and $\frac{1}{12}$?

a. 24

b. 36

c. 48

d. 72

22. $\frac{3}{7} \times \frac{2}{5} =$

a. $\frac{15}{14}$

b. $\frac{1}{6}$

c. $\frac{6}{35}$

d. $\frac{2}{7}$

23. $\frac{5}{12} - \frac{2}{9} =$

a. $\frac{7}{36}$

b. $\frac{1}{3}$

c. $\frac{15}{108}$

d. $\frac{1}{9}$

24. $8\frac{1}{5} - 3\frac{3}{4} + 1\frac{2}{5} - 5\frac{3}{8} =$

a. $-\frac{21}{40}$

b. $1\frac{21}{40}$

c. $\frac{19}{40}$

d. $2\frac{1}{3}$

25. $\frac{5}{7} \div \frac{15}{49} =$

a. $\frac{5}{7}$

b. $\frac{27}{49}$

c. $\frac{3}{7}$

d. $2\frac{1}{3}$

26. $\frac{4}{9} \times \frac{3}{8} =$

a. $\frac{2}{3}$

b. $\frac{1}{4}$

c. $\frac{1}{6}$

d. $\frac{12}{81}$

27. $12\frac{2}{5} \times 3\frac{4}{7} =$

a. $\frac{7}{310}$

b. $36\frac{8}{35}$

c. $44\frac{2}{7}$

d. $52\frac{3}{35}$

28. $13\frac{1}{3} \times \frac{9}{21} \times 2\frac{5}{7} =$

a. $15\frac{25}{49}$

b. $21\frac{33}{76}$

c. $132\frac{5}{21}$

d. $234\frac{45}{441}$

29. Which fraction has the greatest value?

a. $\frac{7}{9}$

b. $\frac{5}{9}$

c. $\frac{15}{18}$

d. $\frac{10}{13}$

30. A box of bricks weighs $22\frac{1}{4}$ pounds. How much will 4 boxes of bricks weigh?

a. $5\frac{9}{16}$ pounds

b. $88\frac{1}{4}$ pounds

c. 89 pounds

d. $92\frac{1}{4}$ pounds

31. What is the product of $\frac{4}{7}$ and $\frac{91}{100}$?

a. $\frac{36}{70}$

b. $\frac{13}{25}$

c. $12\frac{12}{25}$

d. $13\frac{7}{100}$

32. $\frac{3}{7} \div \frac{4}{9} =$

a. $\frac{4}{21}$

b. $1\frac{1}{27}$

c. $\frac{27}{28}$

d. $\frac{12}{63}$

33. $8 \div \frac{2}{5} =$

a. $\frac{16}{5}$

b. 20

c. $3\frac{1}{5}$

d. $\frac{10}{8}$

34. $\frac{17}{35} \div \frac{34}{87} =$

a. $1\frac{17}{70}$

b. $\frac{70}{87}$

c. $\frac{2}{35}$

d. $\frac{2}{87}$

35. $\frac{3}{77} \div \frac{12}{231} =$

a. $1\frac{1}{3}$

b. $\frac{3}{4}$

c. $\frac{21}{77}$

d. $\frac{36}{231}$

36. Divide $2\frac{1}{4}$ by $1\frac{1}{5}$.

a. $2\frac{1}{20}$

b. $2\frac{3}{5}$

c. $\frac{8}{15}$

d. $1\frac{7}{8}$

37. $\frac{\frac{2}{3}}{\frac{4}{9}}$ is equivalent to

a. $1\frac{1}{2}$

b. $\frac{2}{3}$

c. $1\frac{1}{18}$

d. $\frac{5}{9}$

38. $\frac{\frac{13}{30}}{\frac{15}{26}}$ is equivalent to

a. $\frac{1}{4}$

b. $\frac{3}{8}$

c. $\frac{23}{170}$

d. $\frac{169}{225}$

39. Dividing by $\frac{2}{7}$ is the same as

a. multiplying by 7 and dividing by 2

b. multiplying by 2 and dividing by 7

c. multiplying by 7 and multiplying by 2

d. multiplying by $\frac{2}{7}$

40. Which expression can be used to calculate $\frac{1}{5}$ of a \$2,300 bill?

a. $2,300 \div \frac{1}{5}$
b. $2,300 \times \frac{1}{5}$
c. $\frac{2,300}{\frac{1}{5}}$
d. $\frac{\frac{1}{5}}{2,300}$

41. Martin's stock portfolio increased by $\frac{1}{4}$ and then decreased by $\frac{1}{3}$. If his portfolio was worth \$5,400 before, how much is it worth now?

a. \$6,750
b. \$5,400
c. \$4,500
d. \$2,250

42. A barrel was filled $\frac{1}{4}$ of the way with water. Derrick added 18 gallons more, filling the barrel to its capacity. How many gallons are in the barrel now?

a. 20 gallons
b. 22 gallons
c. 24 gallons
d. 28 gallons

43. What is $\frac{2}{3}$ of 48,000?

a. 32,000
b. 36,000
c. 40,000
d. 42,000

44. One plank of wood is $18\frac{13}{16}$ inches long and another is $11\frac{1}{4}$ inches long. What is their combined length in inches?

a. $1\frac{1}{16}$
b. $29\frac{14}{20}$
c. $29\frac{1}{16}$
d. $30\frac{1}{16}$

45. If a delivery of screws is $5\frac{1}{2}$ gross, how many screws are there? Note: 1 gross = 144 units.

a. 360
b. 648
c. 720
d. 792

46. Jared had to sort 400 referrals into the appropriate folders. In the first hour he sorted $\frac{1}{4}$ of the total. In the second hour he sorted $\frac{2}{5}$ of the remainder. How many referrals does he still have to sort?

a. 100
b. 120
c. 180
d. 200

47. A large bag of pebbles weighs $12\frac{1}{4}$ pounds. How many quarter-pound bags of pebbles can be made from this large bag?

a. exactly 3 bags
b. three bags with some pebbles remaining
c. twenty-four bags with some pebbles remaining
d. exactly 49 bags

48. Greg earned one-quarter of his annual income by working as a freelancer. If he made 32,000 dollars this year, how much did he make freelancing?

a. $4,000
b. $8,000
c. $12,000
d. $16,000

49. JoAnne gave $\frac{3}{8}$ of her savings to Karl. If JoAnne initially had $600, how much did JoAnne give Karl?

a. $180
b. $200
c. $225
d. $240

50. Candice placed wood chips around each of the 8 trees in her yard. If she used $3\frac{1}{2}$ bags of wood chips in all, what fraction of a bag did each tree get?

a. $2\frac{2}{7}$
b. $\frac{7}{16}$
c. $\frac{3}{8}$
d. $\frac{16}{7}$

ANSWERS

1. **c.** To find the LCD you should first find the prime factorization of each denominator:
$12 = 3 \times 4 = 3 \times 2 \times 2$
$9 = 3 \times 3$
$18 = 9 \times 3 = 3 \times 3 \times 3$
$24 = 6 \times 4 = 2 \times 3 \times 2 \times 2$
Next, consider the prime factorization of the LCD which must have all of the prime numbers in all of the original denominators.
$12 = 3 \times 2 \times 2$ .... The LCD must have 1 three and 2 twos
$9 = 3 \times 3$ .... The LCD must have 2 threes.
$18 = 3 \times 3 \times 3$ .... The LCD must have 3 threes
$24 = 2 \times 3 \times 2 \times 2$ .... The LCD must have 3 twos
By multiplying 3 threes and 2 twos, the new denominator will be divisible by all of the old denominators. $3 \times 3 \times 3 \times 2 \times 2 \times 2 = 216$.

2. **b.** To find the sum of fractions with common denominators (bottoms), you simply add the two numerators (tops) and keep the same denominator. $\frac{2}{9} + \frac{5}{9} = \frac{7}{9}$.

3. **c.** To change $3\frac{4}{5}$ into an improper fraction, multiply 3×5 plus 4 and place this value over 5:
$\frac{(3 \times 5) + 4}{5} = \frac{15 + 4}{5} = \frac{19}{5}$
Note that the whole number is multiplied by the denominator and added to the numerator to make the new numerator. The same denominator is kept.

4. **b.** $\frac{5}{6} - \frac{1}{6} = \frac{4}{6}$. Dividing top and bottom by 2, this reduces to $\frac{2}{3}$.

5. **a.** $29 \div 3 = 9$ with a remainder of 2. Since the original value was represented as thirds, you put the remainder over 3. The answer is then $9\frac{2}{3}$.

6. **c.** First, we will find the LCD (least common denominator). In this case, it is $9 \times 4 = 36$. Converting, we get $(\frac{4}{9})(\frac{4}{4}) + (\frac{3}{4})(\frac{9}{9}) = \frac{16}{36} + \frac{27}{36} = \frac{43}{36} = 1\frac{7}{36}$.

7. **a.** First, change $1\frac{2}{5}$ into an improper fraction. We multiply 1×5 plus 2 and place this value over 5: $\frac{(1 \times 5) + 2}{5} = \frac{5 + 2}{5} = \frac{7}{5}$.
The expression is now $\frac{7}{5} + \frac{2}{9}$. Next, find the least common denominator (LCD). In this case 9×5, or 45 is the LCD. Converting, we get: $(\frac{7}{5})(\frac{9}{9}) + (\frac{2}{9})(\frac{5}{5}) = \frac{63}{45} + \frac{10}{45} = \frac{73}{45}$. Divide 73 by 45 to get the mixed number $1\frac{28}{45}$.

8. **b.** To reduce $\frac{9}{54}$ to simplest form, just divide the top and bottom by 9 to yield $\frac{1}{6}$.

9. **b.** To change the improper fraction $\frac{154}{11}$ to a whole number, divide 154 by 11: $154 \div 11 = 14$.

10. **c.** You can add all of the whole number parts first: $15 + 9 + 7 + 23 = 54$. Next, add up the fractional parts: $\frac{1}{4} + \frac{2}{3} + \frac{1}{5} + \frac{1}{2}$. Here the LCD is 60; converting, we have $\frac{15}{60} + \frac{40}{60} + \frac{12}{60} + \frac{30}{60}$; adding yields $\frac{97}{60} = 1\frac{37}{60}$; add this to the 54 to get $55\frac{37}{60}$.

11. d. The term *improper fraction* is used to describe a fraction whose top part (numerator) is larger than its bottom part (denominator). To convert $8\frac{2}{3}$ to an improper fraction you multiply the whole number part by the denominator, add the numerator and then put this number over the original denominator. Here, first calculate 8 × 3, add 2, and then put this number over 3: $\frac{(8 \times 3) + 2}{3} = \frac{24 + 2}{3} = \frac{26}{3}$

12. b. To change the improper fraction $\frac{20}{4}$ to a whole number, simply divide 20 by 4: 20 ÷ 4 = 5.

13. a. Multiply the whole number, 2, by the denominator and add the numerator. Then, write this value over 8 (the initial denominator). $2\frac{5}{8}$ = 2 times 8 plus 5 over 8 = $\frac{(2 \times 8) + 5}{8} = \frac{16 + 5}{8} = \frac{21}{8}$.

14. d. First, we convert $1\frac{1}{3}$ to an improper fraction by multiplying the whole number, 1, by the denominator and adding the numerator. $1\frac{1}{3}$ = 1 times 3 plus 1 over 3 = $\frac{(1 \times 3) + 1}{3} = \frac{3 + 1}{3} = \frac{4}{3}$
To take the reciprocal of $\frac{4}{3}$, we just switch the numerator with the denominator. So, the reciprocal of $\frac{4}{3}$ is $\frac{3}{4}$.

15. b. The term *improper fraction* is used to describe a fraction whose top part (numerator) is larger than its bottom part (denominator). Only choice **b** fits this description.

16. d. You can convert the fractions into twentieths in order to perform subtraction. Thus, $22\frac{1}{4} - 13\frac{1}{5} = 22\frac{5}{20} - 13\frac{4}{20} = 9\frac{1}{20}$.

17. d. Only two of the choices are greater than $\frac{1}{2}$, choices **b** and **d**. You can compare the two choices by converting choice **d**, $\frac{3}{4}$ into sixteenths by multiplying top and bottom by 4. Thus, $\frac{3}{4} = \frac{3 \times 4}{4 \times 4} = \frac{12}{16}$. This is greater than choice **b**, $\frac{10}{16}$, and is thus the largest fraction present.

18. b. Convert $18\frac{1}{5}$ to an improper fraction and then multiply by 35.
$\frac{91}{5} \times 35 = \frac{91}{5} \times \frac{35}{1} = \frac{3,185}{5} = 637$.

19. c. First, find the prime factorization of each denominator present:
3 = 3
4 = 2 × 2
16 = 2 × 2 × 2 × 2
Next, make the prime factorization of your new denominator, making sure it contains the prime factorization of all the old denominators.
3 = 3 . . . . LCD must have 1 three.
4 = 2 × 2 . . . . LCD must have 2 twos.
16 = 2 × 2 × 2 × 2 . . . . LCD must have 4 twos.
Thus, the LCD will be 3 × 2 × 2 × 2 × 2 = 48.

20. a. $\frac{23}{5}$ can be converted into a mixed number by dividing 23 by 5 and putting the remainder over 5: 23 ÷ 5 = 4 with a remainder of 3 = $4\frac{3}{5}$.

21. **b.** First, find the prime factorization of each denominator present:
$18 = 6 \times 3 = 2 \times 3 \times 3$
$4 = 2 \times 2$
$12 = 6 \times 2 = 2 \times 3 \times 2$
Next, make the prime factorization of your new denominator, making sure it contains the prime factorization of all the old denominators.
$18 = 2 \times 3 \times 3$ ...... LCD must have 1 two and 2 threes.
$4 = 2 \times 2$ .... LCD must have 2 twos.
$12 = 2 \times 3 \times 2$ ...... LCD must have 2 twos and a 3.
Thus the LCD will be $3 \times 3 \times 2 \times 2 = 36$.

22. **c.** When multiplying fractions, just multiply numerator × numerator and denominator × denominator: $\frac{3}{7} \times \frac{2}{5} = \frac{6}{35}$.

23. **a.** First, find the LCD. $12 = 3 \times 2 \times 2$ and $9 = 3 \times 3$, so the LCD $= 3 \times 3 \times 2 \times 2 = 9 \times 4 = 36$. $\frac{5}{12} - \frac{2}{9} = \frac{15}{36} - \frac{8}{36} = \frac{7}{36}$.

24. **c.** We can first combine all of the whole number portions of each term. $8 - 3 + 1 - 5 = 1$. Next, we will combine the fractional parts. Be careful to take note of the signs: $\frac{1}{5} - \frac{3}{4} + \frac{2}{5} - \frac{3}{8}$. Here the LCD is 40; converting, we have $\frac{8}{40} - \frac{30}{40} + \frac{16}{40} - \frac{15}{40} = -\frac{21}{40}$; last we combine the whole number part with the fractional part: $1 - \frac{21}{40} = \frac{40}{40} - \frac{21}{40} = \frac{19}{40}$.

25. **d.** Note that dividing by $\frac{15}{49}$ is the same as multiplying by $\frac{49}{15}$. Rewrite this question as a multiplication problem, reduce, and solve as follows:

$$\frac{5}{7} \div \frac{15}{49} = \frac{\overset{1}{\cancel{5}}}{\underset{1}{\cancel{7}}} \times \frac{\overset{7}{\cancel{49}}}{\underset{3}{\cancel{15}}} = \frac{1}{1} \times \frac{7}{3} = \frac{7}{3} = 2\frac{1}{3}$$

26. **c.** You can do some canceling in this multiplication problem:

$$\frac{\overset{1}{\cancel{4}}}{\underset{3}{\cancel{9}}} \times \frac{\overset{1}{\cancel{3}}}{\underset{2}{\cancel{8}}} = \frac{1}{3} \times \frac{1}{2} = \frac{1 \times 1}{3 \times 2} = \frac{1}{6}$$

27. **c.** First, convert these mixed numbers into improper fractions: $12\frac{2}{5} = \frac{12 \times 5 + 2}{5} = \frac{62}{5}$. $3\frac{4}{7} = \frac{3 \times 7 + 4}{7} = \frac{25}{7}$. Next, multiply: $\frac{62}{5} \times \frac{25}{7}$. You can cancel before proceeding:

$$\frac{62}{\underset{1}{\cancel{5}}} \times \frac{\overset{5}{\cancel{25}}}{7} = \frac{310}{7} = 44\frac{2}{7}$$

28. **a.** First, you should convert the mixed numbers into improper fractions. $13\frac{1}{3} = \frac{13 \times 3 + 1}{3} = \frac{40}{3}$ and $2\frac{5}{7} = \frac{2 \times 7 + 5}{7} = \frac{19}{7}$. Next, multiply all three numbers, reducing wherever possible:

$$\frac{40}{\underset{3}{\cancel{3}}} \times \frac{\overset{3}{\cancel{9}}}{21} \times \frac{19}{7} = \frac{40}{1} \times \frac{\overset{1}{\cancel{3}}}{\underset{7}{\cancel{21}}} \times \frac{19}{7} = \frac{760}{49} = 15\frac{25}{49}$$

29. **c.** First of all, you know that choice **a** is greater than choice **b** because $\frac{7}{9} > \frac{5}{9}$. Next by multiplying the $\frac{7}{9}$ by $\frac{2}{2}$ you get $\frac{14}{18}$, which is less than choice **c**. So, all you have to do is compare choices **c** and **d**. You can do this by creating a common denominator: just multiply $18 \times 13 = 234$. $\frac{15}{18} = \frac{195}{234}$ and $\frac{10}{13} = \frac{180}{234}$. Since $\frac{195}{234} > \frac{180}{234}$, choice **c** is the greatest fraction here.

30. **c.** This is a multiplication problem. You just multiply the weight of 1 box by 4 to get the weight of 4 boxes: $22\frac{1}{4} \times 4 = \frac{89}{4} \times 4 = 89$ pounds.

31. **b.** To find the product, you just multiply: $\frac{4}{7} \times \frac{91}{100} =$

$$\frac{\cancel{4}^{1}}{\cancel{7}_{1}} \times \frac{\cancel{91}^{13}}{\cancel{100}_{25}} = \frac{13}{25}$$

32. **c.** Change the division problem into a multiplication problem by flipping the second fraction: $\frac{3}{7} \div \frac{4}{9} = \frac{3}{7} \times \frac{9}{4} = \frac{27}{28}$.

33. **b.** When dividing fractions, we actually change the problem into a multiplication problem. The original problem, $8 \div \frac{2}{5}$, can be written as $\frac{8}{1} \times \frac{5}{2}$, which equals $\frac{4}{1} \times \frac{5}{1} = 4 \times 5 = 20$.

34. **a.** To rewrite this question as a multiplication problem, you multiply the first fraction by the reciprocal of the second fraction. Thus, $\frac{17}{35} \div \frac{34}{87} = \frac{17}{35} \times \frac{87}{34} = \frac{\cancel{17}^{1}}{35} \times \frac{87}{\cancel{34}_{2}} = \frac{87}{70} = 1\frac{17}{70}$.

35. **b.** $\frac{3}{77} \div \frac{12}{231} = \frac{3}{77} \times \frac{231}{12}$. This can be reduced:

$$\frac{\cancel{3}^{1}}{\cancel{77}_{1}} \times \frac{\cancel{231}^{3}}{\cancel{12}_{4}} = \frac{3}{4}$$

36. **d.** First, convert $2\frac{1}{4}$ and $1\frac{1}{5}$ to improper fractions. $2\frac{1}{4} = \frac{9}{4}$ and $1\frac{1}{5} = \frac{6}{5}$. Next, set up the division problem: $\frac{9}{4} \div \frac{6}{5}$. Finally, rewrite this division problem as a multiplication problem by taking the reciprocal of the second fraction: $\frac{9}{4} \times \frac{5}{6} = \frac{3}{4} \times \frac{5}{2} = \frac{15}{8} = 1\frac{7}{8}$.

37. **a.** $\frac{\frac{2}{3}}{\frac{4}{9}}$ can be rewritten as $\frac{2}{3} \div \frac{4}{9}$. Next, dividing by $\frac{4}{9}$ is the same as multiplying by $\frac{9}{4}$: $\frac{2}{3} \times \frac{9}{4} = \frac{18}{12}$. Finally, convert to a mixed number and reduce: $1\frac{6}{12} = 1\frac{1}{2}$.

38. **d.** $\frac{\frac{13}{30}}{\frac{15}{26}}$ is the same as $\frac{13}{30} \div \frac{15}{26}$. Flip the second fraction and change the ÷ sign to a ×:

$\frac{13}{30} \times \frac{26}{15} = \frac{169}{225}$.

39. **a.** Dividing by $\frac{2}{7}$ is the same as multiplying by the reciprocal of $\frac{2}{7}$, which is $\frac{7}{2}$. Multiplying by $\frac{7}{2}$ is the same as multiplying by 7 and dividing by 2.

40. **b.** $\frac{1}{5}$ of a number is $\frac{1}{5}$ times a number. Thus $\frac{1}{5}$ of 2,300 would be $\frac{1}{5} \times 2{,}300$. Choice **b**, $2{,}300 \times \frac{1}{5}$ is equivalent to this.

41. **c.** An increase of $\frac{1}{4}$ of 5,400 would be $\frac{1}{4} \times 5{,}400 = \$1{,}350$. The resulting worth would be 5,400 + 1,350 = \$6,750. A decrease of $\frac{1}{3}$, would be $\frac{1}{3} \times 6{,}750$ = a \$2,250 decrease. The amount left would be \$6,750 – \$2,250 = \$4,500.

42. **c.** 18 gallons represents $\frac{3}{4}$ of the whole capacity. 18 = 6 + 6 + 6, so each quarter is 6 gallons. Four quarters would then add to 24 gallons.

43. **a.** In order to find $\frac{2}{3}$ of 48,000, you just multiply: $\frac{2}{3} \times 48{,}000 = 16{,}000$

$$\frac{2}{\cancel{3}_1} \times \cancel{48{,}000}^{16{,}000} = 32{,}000$$

44. **d.** This is a simple addition problem: $18\frac{13}{16} + 11\frac{1}{4}$. You can add the whole numbers first: $18 + 11 = 29$. In order to add $\frac{13}{16}$ and $\frac{1}{4}$, you need to convert the fourths to sixteenths: $\frac{1}{4} \times \frac{4}{4} = \frac{4}{16}$. Add the fractions together: $\frac{13}{16} + \frac{4}{16} = \frac{17}{16} = 1\frac{1}{16}$. Add $1\frac{1}{16}$ to the 29 to get $30\frac{1}{16}$.

45. **d.** There are 144 screws per 1 gross. This can be written as $144\ \frac{\textit{screws}}{\textit{gross}}$. Multiply: $5\frac{1}{2}$ gross $\times$ $144\ \frac{\textit{screws}}{\textit{gross}} = \frac{11}{2} \times 144 = 11 \times 72 = 792$.

46. **c.** Jared starts with 400 referrals. During the first hour he sorts $\frac{1}{4}$ of the 400: $\frac{1}{4} \times 400 = 100$. He thus has $400 - 100 = 300$ left to sort. In the second hour he sorts $\frac{2}{5}$ of the remaining 300. $\frac{2}{5} \times 300 = 120$ sorted in the second hour. Therefore, he now has $300 - 120 = 180$ referrals left to sort.

47. **d.** You need to divide the large bag into $\frac{1}{4}$-pound bags. Hence, you divide $12\frac{1}{4}$ by $\frac{1}{4}$. First, convert $12\frac{1}{4}$ to the improper fraction $\frac{49}{4}$. Next, set up your problem: $\frac{49}{4} \div \frac{1}{4}$. Convert this division problem into a multiplication problem by flipping the second fraction: $\frac{49}{4} \times \frac{4}{1} = 49$.

48. **b.** To find $\frac{1}{4}$ of his income you multiply 32,000 by $\frac{1}{4}$: $32{,}000 \times \frac{1}{4} = \frac{32{,}000}{4} = 8{,}000$.

49. **c.** JoAnne gave Karl $\frac{3}{8}$ of her \$600. To find $\frac{3}{8}$ of this amount, you just multiply: $\frac{3}{8} \times 600 = 225$.

50. **b.** You just divide the $3\frac{1}{2}$ bags by 8 trees. Since $3\frac{1}{2} = \frac{7}{2}$, you can write the equation as $\frac{7}{2} \div \frac{8}{1} = \frac{7}{2} \times \frac{1}{8} = \frac{7}{16}$.

Decimals

WHAT IS A DECIMAL?

A decimal is a special kind of fraction. You use decimals every day when you deal with money—$10.35 is a decimal that represents ten dollars and 35¢. The decimal point separates the dollars from the cents. Because there are 100¢ in one dollar, 1¢ is $\frac{1}{100}$ of a dollar, or $.01.

Each decimal digit to the right of the decimal point has a name:

Examples:

.1 = 1 tenth = $\frac{1}{10}$

.02 = 2 hundredths = $\frac{2}{100}$

.003 = 3 thousandths = $\frac{3}{1,000}$

.0004 = 4 ten-thousandths = $\frac{4}{10,000}$

When you add zeros after the rightmost decimal place, you don't change the value of the decimal. For example, 6.17 is the same as all of these:

6.170
6.1700
6.17000000000000000

If there are digits on both sides of the decimal point (like 10.35), the number is called a *mixed decimal.* If there are digits only to the right of the decimal point (like .53), the number is called a decimal. A whole number (like 15) is understood to have a decimal point at its right (15.). Thus, 15 is the same as 15.0, 15.00, 15.000, and so on.

▶ CHANGING FRACTIONS TO DECIMALS

To change a fraction to a decimal, divide the bottom number into the top number after you put a decimal point and a few zeros on the right of the top number. When you divide, bring the decimal point up into your answer.

Example: Change $\frac{3}{4}$ to a decimal.

1. Add a decimal point and 2 zeros to the top number (3): 3.00
2. Divide the bottom number (4) into 3.00: $4\overline{)3.00}$ = .75
   (Be sure to bring the decimal point up into the answer.)
3. The quotient (result of the division) is the answer: .75

Some fractions may require you to add many decimal zeroes in order for the division to come out evenly. In fact, when you convert a fraction like $\frac{2}{3}$ to a decimal, you can keep adding decimal zeroes to the top number forever because the division will never come out evenly! As you divide 3 into 2, you will keep getting 6's:

$2 \div 3 = .6666666666$ etc.

This is called a repeating decimal and it can be written as $.66\overline{6}$. You can approximate it as .67, .667, .6667, and so on.

▶ CHANGING DECIMALS TO FRACTIONS

To change a decimal to a fraction, write the digits of the decimal as the top number of a fraction, and write the decimal's name as the bottom number of the fraction. Then, reduce the fraction, if possible.

Example: .018

1. Write 18 as the top of the fraction: $\underline{18}$
2. Three places to the right of the decimal means thousandths, so write 1,000 as the bottom number: $\frac{18}{1{,}000}$
3. Reduce the top and bottom numbers by 2: $\frac{18 \div 2}{1{,}000 \div 2} = \frac{9}{500}$

Sample Question:

2.47 is equivalent to which fraction below?

a. $\frac{47}{100}$

b. $2\frac{47}{1{,}000}$

c. $2\frac{47}{50}$

d. $2\frac{47}{100}$

2.47 is 2 and 47 ***hundredths***. This is the same as $2\frac{47}{100}$, choice **d.**

▶ COMPARING DECIMALS

Because decimals are easier to compare when they have the same number of digits after the decimal point, tack zeros onto the end of the shorter decimals. Then, all you have to do is compare the numbers as if the decimal points weren't there:

Example: Compare .08 and .1

1. Tack one zero at the end of .1 to get .10.
2. To compare .10 to .08, just compare 10 to 8.
3. Since 10 is larger than 8, .1 is larger than .08.

▶ ADDING AND SUBTRACTING DECIMALS

To add or subtract decimals, line them up so their decimal points are even. You may want to tack on zeros at the end of shorter decimals so you can keep all your digits lined up evenly. Remember, if a number doesn't have a decimal point, then put one at the right end of the number.

Example: 1.23 + 57 + .038

1. Line up the numbers like this:
2. Add

$$\begin{array}{r} 1.230 \\ 57.000 \\ \underline{+\ .038} \\ 58.268 \end{array}$$

Example: 1.23 – .038

1. Line up the numbers like this:
2. Subtract:

$$\begin{array}{r} 1.230 \\ \underline{-\ .038} \\ 1.192 \end{array}$$

MULTIPLYING DECIMALS

To multiply decimals, ignore the decimal points and just multiply the numbers. Then count the total number of decimal digits (the digits to the right of the decimal point) in the numbers you are multiplying. Count off that number of digits in your answer beginning at the right side and put the decimal point to the left of those digits.

Example: 215.7 × 2.4

1. Multiply 2,157 times 24:

$$\begin{array}{r} 2{,}157 \\ \times\ 24 \\ \underline{51{,}768} \end{array}$$

2. Because there are a total of 2 decimal digits in 215.7 and 2.4, count off 2 places from the right in 51,768, placing the decimal point to the left of the last 2 digits: 517.68

If your answer doesn't have enough digits, tack zeros onto the left of the answer.

Example: .03 × .006

1. Multiply 3 times 6: 3 × 6 = 18
2. You need 5 decimal digits in your answer, so tack on 3 zeroes: 00018
3. Put the decimal point at the front of the number (which is 5 digits in from the right): .00018

▶ DIVIDING DECIMALS

To divide a decimal by a whole number, set up the division (8$\overline{).256}$) and immediately bring the decimal point straight up into the answer. Then, divide as you would normally divide whole numbers:

Example:

```
   .032
8).256
 - 24
    16
  - 16
     0
```

To divide any number by a decimal, there is an extra step to perform before you can divide. Move the decimal point to the very right of the number you are dividing by, counting the number of places you are moving it. Then, move the decimal point the same number of places to the right in the number you are dividing into. In other words, first change the problem to one in which you are dividing by a whole number.

Example: .06$\overline{)1.218}$

1. Because there are 2 decimal digits in .06, move the decimal point 2 places to the right in both numbers and move the decimal point straight up into the answer:

```
          .
06.)121.8
```

2. Divide using the new numbers:

```
      20.3
06.)121.8
   - 12
      01
     - 0
      18
    - 18
       0
```

Under the following conditions, you have to tack on zeros to the right of the last decimal digit in the number you are dividing into:

- ▶ if there aren't enough digits for you to move the decimal point to the right.
- ▶ if the answer doesn't come out evenly when you do the division.
- ▶ if you are dividing a whole number by a decimal.

PRACTICE QUESTIONS

1. Which of the following choices has a 6 in the tenths place?
- a. 60.17
- b. 76.01
- c. 1.67
- d. 7.061

2. Which of the following choices has a 3 in the hundredths place?
- a. 354.01
- b. .54031
- c. .54301
- d. .03514

3. 234.816 when rounded to the nearest hundredth is
- a. 200
- b. 234.8
- c. 234.83
- d. 234.82

4. Which of the decimals below has the *greatest* value?
- a. .03
- b. .003
- c. .031
- d. .0031

5. 25.682 rounded to the nearest tenth is
- a. 26
- b. 25
- c. 25.68
- d. 25.7

6. What is 3.133 when rounded to the nearest tenth?
- a. 3
- b. 3.1
- c. 3.2
- d. 3.13

7. $\frac{3}{20}$ is equivalent to which of the following decimals?
- a. .03
- b. .06
- c. .60
- d. .15

8. Which number sentence is true?

a. $.23 \geq 2.3$
b. $.023 \leq .23$
c. $.023 \leq .0023$
d. $.023 \geq 2.3$

9. Which decimal below is the smallest?

a. .00782
b. .00278
c. .2780
d. 0.000782

10. Which decimal is equivalent to the fraction $\frac{7}{25}$?

a. .07
b. .35
c. .28
d. .725

11. What is the sum of 8.514 and 4.821?

a. 12.335
b. 13.335
c. 12.235
d. 13.235

12. What is the sum of 2.523 and 6.76014?

a. 9.3
b. 92.8314
c. 9.28314
d. 928.314

13. $67.104 + 51.406 =$

a. 11.851
b. 1185.1
c. 118.51
d. 118.61

14. What is the sum of 3.75, 12.05, and 4.2?

a. 20
b. 19.95
c. 19.00
d. 19.75

15. 14.02 + .987 + 0.145 =

a. 14.152
b. 15.152
c. 14.142
d. 15.142

16. 5.25 + 15.007 + .87436 =

a. 211.3136
b. 20.13136
c. 201.3136
d. 21.13136

17. $\frac{1}{5}$ + .25 + $\frac{1}{8}$ + .409 =

a. $\frac{1}{13}$ + .659
b. .659 + $\frac{1}{40}$
c. .984
d. 1.084

18. What is the sum of 12.05, 252.11, 7.626, 240, and 8.003?

a. 5,197.86
b. 519.789
c. 518.685
d. 518.786

19. What is the sum of –8.3 and 9?

a. 17.3
b. 0.7
c. 1.73
d. 7

20. The following is a list of the thickness of four boards: .52 in, .81 in, .72 in, and 2.03 in. If all four boards are stacked on top of one another, what will the total thickness be?

a. 40.8 in.
b. 0.408 in.
c. 4.008 in.
d. 4.08 in.

21. 324.0073 – 87.663 =

a. 411.6703
b. 236.3443
c. 2363.443
d. 23.634443

22. 8.3 – 1.725 =

a. 6.575
b. 6.775
c. 7.575
d. 10.025

23. 12.125 – 3.44 =

a. 9.685
b. 8.785
c. 8.685
d. 8.585

24. 89.037 – 27.0002 – 4.02

a. 62.0368
b. 59.0168
c. 58.168
d. 58.0168

25. .89735 – .20002 – .11733 =

a. .69733
b. .59733
c. .58033
d. .58

26. What is 287.78 – .782 when rounded to the nearest hundred?

a. 286.998
b. 286.90
c. 286.99
d. 300

27. .0325 – (-.0235) =

a. 0
b. .0560
c. .0650
d. .560

28. .667 – (–.02) – .069 =

a. .618
b. .669
c. .596
d. .06

29. $-12.3 - (-4.2) =$

a. -8.1
b. -16.5
c. 16.5
d. 8.1

30. $-6.5 - 8.32 =$

a. 14.82
b. 1.82
c. $-.82$
d. -14.82

31. $.205 \times .11 =$

a. $.02255$
b. 2255
c. 2.255
d. 22.55

32. $.88 \times .22 =$

a. $.01936$
b. $.1936$
c. $.1616$
d. 1.616

33. $8.03 \times 3.2 =$

a. 24.06
b. 24.6
c. 25.696
d. 156.96

34. $.56 \times .03 =$

a. 168
b. 16.8
c. $.168$
d. $.0168$

35. $.32 \times .04 =$

a. $.128$
b. $.0128$
c. 128
d. 12.8

36. What is the product of 5.49 and .02?

a. .1098
b. 5.51
c. 5.47
d. 274.5

37. .125 × .8 × .32 =

a. .32
b. $\frac{1}{10}$
c. $\frac{8}{250}$
d. $\frac{32}{100}$

38. .15 × $\frac{1}{5}$ =

a. .2
b. .3
c. .02
d. .03

39. If each capsule contains .03 grams of active ingredients, how many grams of active ingredients are in 380 capsules?

a. $126\frac{2}{3}$ grams
b. 11.4 grams
c. 12.6 grams
d. 1.14 grams

40. If a piece of foil is .032 cm. thick, how thick would a stack of 200 such pieces of foil be?

a. 64 cm.
b. 16 cm.
c. 6.4 cm.
d. 1.6 cm.

41. 3.26 ÷ .02 =

a. 163
b. 65.2
c. 16.3
d. 652

42. 512 ÷ .256 =

a. 20
b. 2,000
c. 200
d. 2

43. 3.4 ÷ .17 =

a. 3
b. 2
c. 30
d. 20

44. What is the quotient of 83.4 ÷ 2.1 when rounded to the nearest tenth?

a. 40
b. 39.71
c. 39.7
d. 39.8

45. .895 ÷ .005 =

a. .0079
b. .179
c. 179
d. 1790

46. What is the quotient of .962 ÷ .023 when rounded to the nearest hundredth?

a. 41.83
b. 41.826
c. 42
d. 23.9

47. $\frac{8.4}{.09}$ =

a. $93\frac{1}{3}$
b. .0107
c. .756
d. 75.6

48. $\frac{375}{.125}$ =

a. 5,625
b. 3,000
c. 56.25
d. 30

49. A seventy pound bag of cement can be divided into how many smaller bags, each weighing 3.5 pounds?

a. 20
b. 16
c. 10
d. 5

50. Markers will be placed along a roadway at regular .31 kilometer intervals. If the entire roadway is 1.55 kilometers, how many markers will be used?

a. 480.5
b. 50
c. 48.05
d. 5

ANSWERS

1. **c.** The places to the left of the decimal point are (in order): the *tenths place*, the *hundredths place*, *thousandths place*, and so on. You are looking for a 6 in the tenths place, which is the first spot to the right of the decimal point. Only choice **c** has a 6 in this place:

units (ones)	***tenths***	*hundredths*
1.	**6**	7

Note that choice **b** has a 6 in the tens place and NOT the *tenths* place.

2. **d.** The places to the left of the decimal point are (in order): the *tenths place*, the *hundredths place*, *thousandths place*, and so on. You are looking for a 3 in the hundredths place, which is the second spot to the right of the decimal point. Only choice **d** has a 3 in this place:

units (ones)	*tenths*	***hundredths***	*thousandths*	*ten thousandths*	*hundred thousandths*
0.	0	**3**	5	1	4

Note that choice **a** has a 3 in the hundreds place and NOT the *hundredths* place.

3. d. When rounding to the nearest hundredth, you need to truncate (cut short) the number, leaving the last digit in the hundredths place. If the number after the hundredths place is a 5 or higher, you would round up.

hundreds	*tens*	*units (ones)*	*tenths*	***hundredths***	*thousandths*
2	3	4.	8	**1**	6

↑
6 is higher than 5,
so you round the 1 in the
hundredths place up to 2.

Thus, the answer is 234.82, choice **d**.

4. c. Choice **c** has the greatest value, $\frac{31}{1,000}$. The four choices are compared below:

a. .03	$\frac{3}{100} = \frac{30}{1,000}$
b. .003	$\frac{3}{1,000}$
c. .031	$\frac{31}{1,000}$
d. .0031	$\frac{31}{10,000}$

5. d. 25.682 has a 6 in the tenths place. Because the number in the hundredths place (the 8) is greater than 5, you will round up to 25.7.

tens	*units (ones)*	***tenths***	*hundredths*	*thousandths*
2	5.	**6**	8	2

↑
You round up because 8 ≥ 5.

6. **b.** In order to round to the nearest tenth, you need to cut the number short, leaving the last digit in the tenths place. Here you cut the number short without rounding up because the number in the hundredths place is not ≥ 5.

units (ones)	*tenths*	*hundredths*	*thousandths*
3.	1	3	3

↑ (under the hundredths column)

You don't round up because 3 is less than 5. Thus, the answer is 3.1, choice **b.**

7. **d.** $\frac{3}{20}$ can quickly be converted to hundredths by multiplying by $\frac{5}{5}$: $\frac{3}{20} \cdot \frac{5}{5} = \frac{15}{100}$. $\frac{15}{100}$ is the same as 15 *hundredths*, or .15, choice **d.**

8. **b.** .023 equals $\frac{23}{1,000}$ which is less than .23, which equals $\frac{23}{100}$. Thus $.023 \leq .23$. The symbol $\leq$ means *less than or equal to.*

9. **d.** Each answer choice is equivalent to the values listed below:

choice **a**: $.00782 = \frac{782}{100,000} = \frac{7,820}{1,000,000}$

choice **b**: $.00278 = \frac{278}{100,000} = \frac{2,780}{1,000,000}$

choice **c**: $.2780 = \frac{2,780}{10,000} = \frac{287,000}{1,000,000}$

choice **d**: $.000782 = \frac{782}{1,000,000}$

Thus, choice **d** is the smallest number listed among the choices.

10. **c.** $\frac{7}{25}$ can be translated into *hundredths* by multiplying by $\frac{4}{4}$. Thus, $\frac{7}{25} \times \frac{4}{4} = \frac{28}{100}$. 28 *hundredths* can be rewritten as .28, choice **c.**

11. **b.** *Sum* means *add.* Make sure you line up the decimal points and then add:

$$\begin{array}{r} 8.514 \\ +\ 4.821 \\ \hline 13.335 \end{array}$$

12. **c.** *Sum* means *add.* Line up the decimal points and add:

$$\begin{array}{r} 2.523 \\ +\ 6.76014 \\ \hline 9.28314 \end{array}$$

13. **c.** Line up the decimal points and add:

$$\begin{array}{r} 67.104 \\ +\ 51.406 \\ \hline 118.510 \end{array}$$

14. **a.** 4.2 is equivalent to 4.20. Line up all the decimal points and add:

$$\begin{array}{r} 3.75 \\ 12.05 \\ +\ 4.20 \\ \hline 20.00 \end{array}$$

15. **b.** 14.02 is equivalent to 14.020. Line up all the decimal points and add:

$$\begin{array}{r} 14.020 \\ .987 \\ +\ .145 \\ \hline 15.152 \end{array}$$

16. **d.** Add zeros as space holders to the numbers 5.25 and 15.007. Then, line all the numbers up by their decimal points and add:

$$\begin{array}{r} 5.25000 \\ 15.00700 \\ +\ .87436 \\ \hline 21.13136 \end{array}$$

17. **c.** First, convert the fractions to decimals: $\frac{1}{5}$ = .2 and $\frac{1}{8}$ = .125. Next, line up all the numbers by their decimal points and add (note that zeros are added as place holders):

$$\begin{array}{r} .200 \\ .250 \\ .125 \\ +\ .409 \\ \hline .984 \end{array}$$

18. **b.** *Sum* signifies addition. Line up the decimal points and add. Note that zeros can be added as place holders:

$$\begin{array}{r} 12.050 \\ 252.110 \\ 7.626 \\ 240.000 \\ +\ 8.003 \\ \hline 519.7890 \end{array}$$

19. **b.** 9 plus -8.3 is the same as 9 minus 8.3. Rewrite 9 as 9.0 and subtract:

$$\begin{array}{r} 9.0 \\ -\ 8.3 \\ \hline .7 \end{array}$$

20. **d.** Line up the decimal points and add:

$$\begin{array}{r} .52 \\ .81 \\ .72 \\ +\ 2.03 \\ \hline 4.08 \end{array}$$

21. **b.** Line up the decimal points and subtract:

$$\begin{array}{r} 324.0073 \\ -\ 87.663 \\ \hline 236.3443 \end{array}$$

22. **a.** Rewrite 8.3 as its equivalent 8.300. Line up the decimal points and subtract:

$$\begin{array}{r} 8.300 \\ -\ 1.725 \\ \hline 6.575 \end{array}$$

23. **c.** Line up the decimal points and subtract:

$$\begin{array}{r} 12.125 \\ -\ 3.44 \\ \hline 8.685 \end{array}$$

24. **d.** First, rewrite 89.037 as its equivalent 89.0370. Next, subtract 27.0002:

$$\begin{array}{r} 89.0370 \\ -\ 27.0002 \\ \hline 62.0368 \end{array}$$

Now, you must subtract 4.02 from the 62.0386: (If you chose choice **a**, you forgot the next step.)

$$\begin{array}{r} 62.0368 \\ -\ 4.02 \\ \hline 58.0168 \end{array}$$

25. **d.** Perform the indicated operations (subtractions) in 2 steps:

$$\begin{array}{r} .89735 \\ -\ .20002 \\ \hline .69733 \end{array}$$

Next, subtract .11733 from .69733 to get .58.

26. **d.** The question asks you to round to the hundred (not *hundredth!*). 287.78 – .782 = 286.998. When this value is rounded to the nearest hundred, you get 300.

27. **b.** Subtracting a negative is the same as adding a positive. Thus, .0325 – (–.0235) is the same as .0325 + .0235. Adding, you get .0560.

28. **a.** Subtracting a negative is the same as adding a positive. Thus, .667 – (–.02) – .069 = .667 + .02 –.069. This equals .687 – .069 = .618.

29. a. Subtracting a negative number is the same as adding a positive number. Thus, –12.3 – (–4.2) = –12.3 + +4.2. – 12.3 + 4.2 will yield a negative value because you are starting 12.3 units away from zero in the *negative* direction. Adding 4.2 will bring you closer to 0, but you will still have a negative answer. To figure out what the answer is, subtract 4.2 from 12.3 and add a minus sign. Thus, you get –8.1.

30. d. –6.5 – 8.32 is the same as –6.5 + –8.32. When adding 2 negative numbers, first ignore the negative signs and add in the normal fashion. 6.5 + 8.32 = 14.82. Next, insert the negative sign to get –14.82, choice **d.**

31. a. First, multiply in the usual fashion (ignoring the decimal points): 205 × 11 = 2,255. Next, you need to insert the decimal point in the correct position, so take note of the position of each decimal point in the two factors:

.532 the decimal point is **3** places to the left.
.89 the decimal point is **2** places to the left.
In the answer, the decimal point should be **3** + **2** , or **5** places to the left.
2,255 becomes .02255, choice **a.**

32. b. First, multiply in the usual fashion (ignoring the decimal points): 88 × 22 = 1,936. Next, you need to insert the decimal point in the correct position, so take note of the position of each decimal point in the two factors:

.88 the decimal point is **2** places to the left.
.22 the decimal point is **2** places to the left.
In the answer, the decimal point should be **2** + **2** , or **4** places to the left.
1,936 becomes .1936, choice **b.**

33. c. First, multiply in the usual fashion (ignoring the decimal points): 803 × 32 = 25,696. Next, you need to insert the decimal point in the correct position, so take note of the position of each decimal point in the two factors:

8.03 the decimal point is **2** places to the left.
3.2 the decimal point is **1** place to the left.
In the answer, the decimal point should be **3** places to the left.
25,696 becomes 25.696, choice **c.**

34. d. Multiply in the usual fashion, and insert the decimal point 4 places to the left:

.56 the decimal point is **2** places to the left.
.03 the decimal point is **2** places to the left.
In the answer, the decimal point should be **4** places to the left.
56 × 03 = 168 (when ignoring decimal) and becomes .0168 when you insert the decimal point 4 places to the left. Thus, the answer is choice **d.**

35. b. Multiply in the usual fashion, and insert the decimal point 4 places to the left: .32 × .04 = .0128.

36. a. The term *product* signifies multiplication. Multiply 5.49 by .02 in the usual fashion, and insert the decimal point 4 places to the left: 5.49 × .02 = .1098.

37. **c.** First, multiply .125 by .8 to get .1. Next, multiply .1 by .32 to get .032. This answer is equivalent to 32 thousandths, or $\frac{32}{1000}$. This reduces to $\frac{4}{125}$, choice **c.**

38. **d.** First, convert $\frac{1}{5}$ to a decimal: $\frac{1}{5} = 1 \div 5 = .2$. Next, multiply: $.15 \times .2 = .03$

39. **b.** Multiply the amount of active ingredients in one capsule (.03) by the number of capsules (380): $380 \times .03 = 11.4$ grams.

40. **c.** To solve, simply multiply the thickness of each piece by the total number of pieces. $200 \times .032 = 6.4$ cm.

41. **a.** The division problem $3.26 \div .02$ can be solved with long division. First, just move the decimal point 2 places to the right in each number:

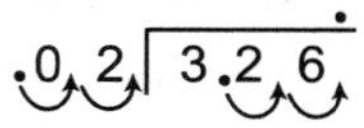

$.02\overline{)3.26}$

Next, divide as usual to get 163, choice **a.**

42. **b.** The division problem $512 \div .256$ can be solved with long division. Move the decimal point 3 places to the right in each number:

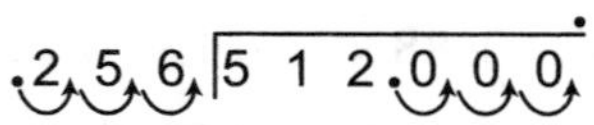

$.256\overline{)512.000}$

Next, divide as usual to get 2,000, choice **b.**

43. **d.** The division problem $3.4 \div .17$ can be solved with long division. First, just move the decimal point 2 places to the right in each number:

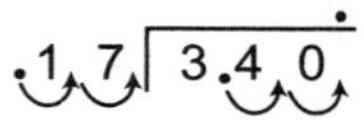

$.17\overline{)3.40}$

Next, divide as usual to get 20, choice **d.**

44. **c.** The division problem $83.4 \div 2.1$ can be solved with long division, moving the decimal point in each number 1 place to the right:

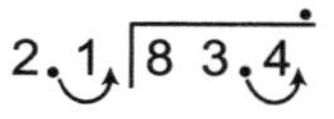

$2.1\overline{)83.4}$

Next, divide as usual to get 39.714286. Finally, round to the nearest tenth: 39.7, choice **c.**

45. **c.** The division problem $.895 \div .005$ can be solved with long division, moving the decimal point in each number 3 places to the right:

$.005\overline{).895}$

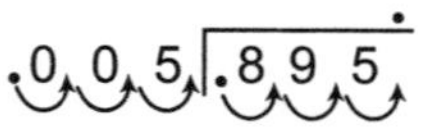

Next, divide to get the answer: 179, choice **c.**

46. **a.** The division problem $.962 \div .023$ can be solved with long division, moving the decimal point in each number 3 places to the right:

$.023\overline{).962}$

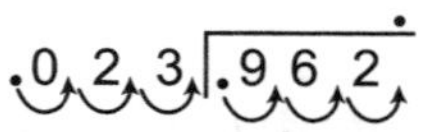

Next, divide to get 41.826087. Rounding this number to the nearest hundredth yields 41.83, choice **a.**

47. a. The division problem 8.4 ÷ .09 can be solved with long division, moving the decimal point in each number 2 places to the right:

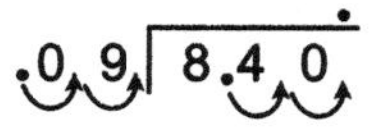

.0 9) 8.4 0.

Dividing yields an answer of 93.333333 . . . or $93\frac{1}{3}$, choice **a.**

48. b. The division problem 375 ÷ .125 can be solved with long division, moving the decimal point in each number 3 places to the right:

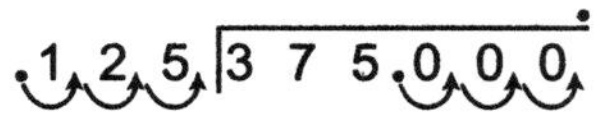

.1 2 5) 3 7 5.0 0 0.

Dividing yields 3,000, choice **b.**

49. a. To solve, divide 70 by 3.5. 70 ÷ 3.5 can be solved with long division, moving the decimal point in each number 1 place to the right:

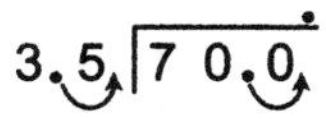

3.5) 7 0.0.

Next, divide as usual to get 20, choice **a.**

50. d. To solve, divide the total 1.55 km distance by the interval, .31 km. 1.55 ÷ .31 can be solved with long division. The decimal point in each number is moved two places to the right:

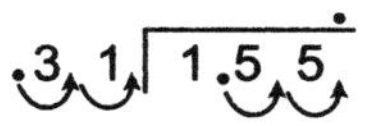

.3 1) 1.5 5.

Next, divide to get the answer: 5, choice **d.**

Number Series and Analogies

NUMBER SERIES

Some number series can be categorized as *arithmetic* or *geometric*. Other number series are neither arithmetic or geometric and, thus, must be analyzed in search of a pattern.

Let's review the two general number series you will see on the test:

1. **Arithmetic Series**

   Arithmetic series progress by adding (or subtracting) a constant number to each term. For example, look at the series:

   4, 7, 10, 13, 16, . . .

   Notice that each term is three more than the term that comes before it. Therefore, this is an arithmetic series with a *common difference* of 3.

2. Geometric Series

Geometric series progress by multiplying each term by a constant number to get the next term. For example, look at the series:

$\frac{1}{2}$, 1, 2, 4, 8, 16, 32, . . .

Notice that each term is two times the prior term. Therefore, this is a geometric series with a *common ratio* of 2.

Sample Question:

What number comes next in the following series?

65, 72, 79, 86, ___

a. 87
b. 89
c. 90
d. 93

This is an arithmetic series with a common difference of 7. Thus, the next number will be 86 + 7 = 93. The correct answer is **d.**

▶ LETTER SERIES

Instead of containing numbers, letter series use the relationship of the letters in the alphabet to generate patterns. Study the series and try to figure out what the relationship is.

For example, look at the series:

ABC CBA DEF FED GHI ____

Which answer choice below will correctly fill in the blank?

a. IJK
b. JKL
c. LKJ
d. IHG

The correct answer is **d.** Notice that the first *triplet* of the series is ABC. The next triplet contains the same three letters listed in reverse order: CBA. The third triplet is DEF, followed by its inverse FED. Next comes GHI, so the missing three letters will be GHI in reverse order, or IHG.

SYMBOL SERIES

Symbol series are visual series based on the relationship between images. Carefully analyze this visual series to find the pattern.

For example, look at the symbol series below:

↑ ↗ → ↘ ↓ ↙ ___

Select the answer choice that best completes the sequence below.

a. ↖
b. ←
c. ↑
d. ↔

Notice that the position of each arrow can be found by rotating the previous arrow by 45° clockwise. Thus, the next arrow will be: ←, choice **b**.

PRACTICE QUESTIONS

1. What number is missing from the series below?
18 14 ___ 6 2
a. 12
b. 10
c. 8
d. 4

2. What number is missing from the series below?
5 15 45 ___ 405
a. 50
b. 60
c. 75
d. 135

3. What number is missing from the series below?
72 67 ___ 57 52
a. 62
b. 63
c. 59
d. 58

4. What number is missing from the series below?

8.2 ___ 7.6 7.3 7.0

a. 8.1
b. 8
c. 7.9
d. 7.8

5. What number is missing from the series below?

1 4 6 1 ___ 6 1

a. 6
b. 4
c. 1
d. 2

6. What number is missing from the series below?

9.7 10.1 ___ 10.9 11.3

a. 9.7
b. 9.9
c. 10.5
d. 11.3

7. What number is missing from the series below?

0 1 8 27 ___

a. 34
b. 54
c. 64
d. 76

8. What number is missing from the series below?

567, 542, 517, 492, . . .

a. 499
b. 483
c. 477
d. 467

9. What number is missing from the series below?

90 45 ___ 11.25 5.625

a. 0
b. 12.5
c. 16
d. 22.5

10. What number is missing from the series below?

___ .34 .068 .0136

a. 1.7
b. .408
c. 4.08
d. 17

11. What number should come next in the series below?

2, 1, $\frac{1}{2}$, $\frac{1}{4}$, . . .

a. $\frac{1}{3}$
b. $\frac{1}{8}$
c. $\frac{2}{8}$
d. $\frac{1}{16}$

12. What number is missing from the series below?

0 1 ___ 6 10 15

a. 2
b. 3
c. 4
d. 5

13. What number is missing from the series below?

4 1 5 4 1 7 4 1 9 4 1 ___

a. 1
b. 4
c. 9
d. 11

14. What number is missing from the series below?

$\frac{2}{5}$ $\frac{1}{15}$ ___ $\frac{1}{540}$ $\frac{1}{3,240}$

a. $\frac{2}{30}$
b. $\frac{1}{45}$
c. $\frac{1}{90}$
d. $\frac{1}{270}$

15. What number is missing from the series below?

30 ___ 27 $25\frac{1}{2}$ 24

a. $29\frac{1}{2}$

b. 29

c. $28\frac{1}{2}$

d. 28

16. What number is missing from the series below?

10 12 16 22 30 40 ___

a. 33

b. 34

c. 40

d. 52

17. What number is missing from the series below?

−12 6 4 −13 7 3 −14 ___ 2

a. 8

b. 10

c. 12

d. 13

18. What number is missing from the series below?

5,423 5,548 5,673 5,798 ___

a. 5,823

b. 5,848

c. 5,923

d. 5,948

19. What number is missing from the series below?

6 11 16 16 21 26 26 ___

a. 16

b. 26

c. 30

d. 31

20. What number is missing from the series below?

10 14 84 88 264 ___

a. 18

b. 188

c. 268

d. 334

21. What number is missing from the series below?

38 20 5 –7 –16 ___

a. –25
b. –22
c. –20
d. –19

22. What number is missing from the series below?

9 8 16 15 ___ 29 58

a. 30
b. 14
c. 9
d. 8

23. What number should fill in the blank in the series below?

53, 53, ___, 40, 27, 27, . . .

a. 14
b. 38
c. 40
d. 51

24. What number should come next in the series below?

0.2, $\frac{1}{5}$, 0.4, $\frac{2}{5}$, 0.8, $\frac{4}{5}$, . . .

a. $\frac{8}{10}$
b. 0.7
c. 1.6
d. 0.16

25. What number should come next in the series below?

1.5, 2.3, 3.1, 3.9, . . .

a. 4.2
b. 4.4
c. 4.7
d. 5.1

26. What number should come next in the series below?

29, 27, 28, 26, 27, 25, . . .

a. 23
b. 24
c. 26
d. 27

27. What number should come next in the series below?
31, 29, 24, 22, 17, . . .
- **a.** 15
- **b.** 14
- **c.** 13
- **d.** 12

28. What number should fill in the blank in the series below?
10, 34, 12, 31, ____, 28, 16, . . .
- **a.** 14
- **b.** 18
- **c.** 30
- **d.** 34

29. What is the missing term in the number pattern below?
240, 120, 60, 30, 15, ____, $3\frac{3}{4}$
- **a.** $7\frac{1}{2}$
- **b.** $9\frac{1}{4}$
- **c.** 10
- **d.** $11\frac{1}{4}$

30. What number should come next in the series below?
3, 4, 7, 8, 11, 12, . . .
- **a.** 7
- **b.** 10
- **c.** 14
- **d.** 15

31. What number should come next in the series below?
1, 4, 9, 5, 17, . . .
- **a.** 6
- **b.** 8
- **c.** 22
- **d.** 25

32. What number should come next in the series below?
$1, \frac{7}{8}, \frac{3}{4}, \frac{5}{8}, \ldots$
- **a.** $\frac{2}{3}$
- **b.** $\frac{1}{2}$
- **c.** $\frac{3}{8}$
- **d.** $\frac{1}{4}$

33. What two numbers should come next in the series below?
8, 22, 12, 16, 22, 20, 24, . . .
a. 28, 32
b. 28, 22
c. 22, 28
d. 22, 26

34. If the pattern $\frac{1}{2}, \frac{1}{4}, \frac{1}{8}, \frac{1}{16}$ . . . is continued, what is the denominator of the tenth term?
a. 64
b. 212
c. 512
d. 1,024

35. What number should come next in the series below?
14, 28, 20, 40, 32, 64, . . .
a. 52
b. 56
c. 96
d. 128

36. What two numbers should come next in the series below?
9, 12, 11, 14, 13, 16, 15, . . .
a. 14, 13
b. 8, 21
c. 14, 17
d. 18, 17

37. What number should come next in the series below?
21, 24, 30, 21, 36, 42, . . .
a. 21
b. 27
c. 42
d. 46

38. What number should come next in the series below?
XX, XVI, XII, VIII, . . .
a. IV
b. V
c. VI
d. III

39. What number should come next in the series below?
J14, L11, N8, P5, . . .
- **a.** Q2
- **b.** Q3
- **c.** R2
- **d.** S2

40. What number should come next in the series below?
VI, 10, V, 11, IV, 12, . . .
- **a.** VII
- **b.** III
- **c.** IX
- **d.** 13

41. Select the answer choice that best completes the sequence below.
JAK KBL LCM MDN _____
- **a.** OEP
- **b.** NEO
- **c.** MEN
- **d.** PFQ

42. Select the answer choice that best completes the sequence below.
QPO NML KJI ____ EDC
- **a.** HGF
- **b.** CAB
- **c.** JKL
- **d.** GHI

43. Select the answer choice that best completes the sequence below.
ELFA GLHA ILJA ______ MLNA
- **a.** OLPA
- **b.** KLMA
- **c.** LLMA
- **d.** KLLA

44. Select the answer choice that best completes the sequence below.
◦○● | ▴△▵ | □ ___
- **a.** □▫
- **b.** □▪
- **c.** △▴
- **d.** ▫■

45. Select the answer choice that best completes the sequence below.

△□△ | □○□ | ○◇○ | ◇▭ ___

a. ◇
b. ▭
c. ○
d. △

46. Select the answer choice that best completes the sequence below.

⇨⇨ | ⇧⇩ | ⇨⇨ | ⇧⇩ | ___

a. ⇩⇩
b. ⇨⇨
c. ⇩⇧
d. ⇦⇦

47. Select the answer choice that best completes the sequence below.

∫⌢ | ⌢⌠ | ∫⌢ | ___

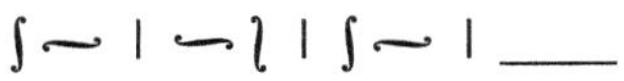

a.  ⌠⌢
b. ⌢⌢
c. ∫⌠
d. 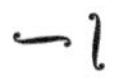 ⌢⌠

48. Select the answer choice that best completes the sequence below.

| ___ ___

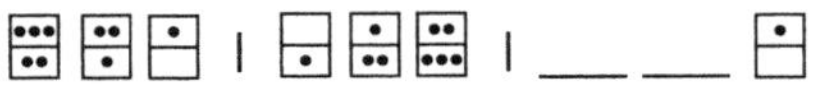

a. 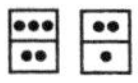
b. 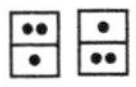
c. 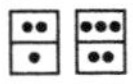
d.

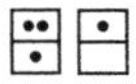

49. Select the answer choice that best completes the sequence below.

E m E | m m m | E ɯ E | ɯ __ ɯ

a. m
b. E
c. ɯ
d. ∃

50. Select the answer choice that best completes the sequence below.

a.

b.

c.

d.

ANSWERS

1. **b.** This is an arithmetic series that decreases by four as the series progresses. Thus, the missing number is 14 – 4 = 10. You can check that this is correct by applying the rule to the 10: 10 – 4 = 6, which is in fact the next term.
2. **d.** This is a geometric series. You multiply each term by 3 to get the next term. The missing term is then 45 × 3 = 135. You can check that this rule works by multiplying 135 by 3. This yields 405, which is the next term.
3. **a.** This is an arithmetic series. Each term is 5 less than the prior term. To find the missing term just subtract 5 from 67 to get 62. Next, check that the rule is correct by verifying 62 – 5 = 57, the next term.
4. **c.** This is an arithmetic series with a common difference of .3. This simply means that each term is .3 less than the term before it. 8.2 – .3 = 7.9, so the missing term is 7.9. To check that you found the right rule, subtract .3 from 7.9 to get 7.3, the next term.
5. **b.** This series is neither arithmetic or geometric. It is simply three numbers repeating over and over in order. The numbers 1, 4, and 6 repeat. Thus, the missing number is 4.
6. **c.** This is an arithmetic series. Each term is .4 greater than the previous term. 10.1 + .4 = 10.5. Using this rule, the term following 10.5 should be 10.5 + .4 = 10.9, and it is. Thus, you know you used the correct rule.
7. **c.** This series is neither arithmetic or geometric. If you look carefully at the numbers, you should notice that each is a cube of a number. In other words, 0, 1, 8, 27 corresponds to 0^3, 1^3, 2^3, 3^3, so the next term should equal 4^3, or 64.
8. **d.** This is an arithmetic series; each number is 25 less than the previous number. Thus, the answer is 492 – 25 = 467.
9. **d.** This is a geometric series with a common ratio of $\frac{1}{2}$. In other words, each term is $\frac{1}{2}$ of the term that precedes it. Thus, the missing term is $\frac{1}{2}$ of 45. $\frac{1}{2} \times 45 = 22.5$. To check that you used the correct rule, take $\frac{1}{2}$ of 22.5: $22.5 \times \frac{1}{2} = 11.25$. This is the next term in the series so you know you are right.
10. **a.** This is a geometric series with a common ratio of .2. In other words, each term is .2 times the term that precedes it. You can divide .34 by .2 to figure out what the first term is. .34 ÷ .2 = 1.7. You can check that you have the correct answer by applying the rule: 1.7 × .2 = .34.
11. **b.** This is a geometric series; each number is one-half of the previous number. Thus, the next number should be $\frac{1}{2} \times \frac{1}{4} = \frac{1}{8}$.
12. **b.** Here, the numbers are increasing, but the amount by which they are increasing is increasing as well. 0 (+ 1) 1 (+2) 3 (+3) 6 (+4) 10 (+5) 10. Thus, the missing number is 3.
13. **d.** Consider this series as a triplet. The first two terms of the triplet are always 4 followed by 1. Notice that every third term gets 2 added to it: 4 1 5 4 1 7 4 1 9 4 1 ___. Thus, the missing number is 9 + 2 = 11.

14. c. This is a geometric series with a common ratio of $\frac{1}{6}$. This means that each term is the prior term multiplied by $\frac{1}{6}$. This is more evident when looking at the last two terms of the series: $\frac{2}{5}$ $(\times \frac{1}{6})$ $\frac{1}{15}$ $(\times \frac{1}{6})$ ___ $(\times \frac{1}{6})$ $\frac{1}{540}$ $(\times \frac{1}{6})$ $\frac{1}{3,240}$. Thus, the missing term is $\frac{1}{15} \times \frac{1}{6} = \frac{1}{90}$.

15. c. This is an arithmetic series with a common difference of $1\frac{1}{2}$. The missing term is $30 - 1\frac{1}{2} = 28\frac{1}{2}$. You can check your work by applying the rule to $28\frac{1}{2}$. $28\frac{1}{2} - 1\frac{1}{2} = 27$, which is the next term.

16. d. Here, the numbers are increasing. Notice that it is not a steady common difference (arithmetic), nor a steady common ratio (geometric). The amount of increase corresponds more to an addition, and each term is increasing by having a larger number added to it. The pattern here is 10 (+2) 12 (+4) 16 (+6) 22 (+8) 30 (+10) 40 (+12) ___. Thus, the missing number is 40 + 12, or 52.

17. a. Here, the series can be considered as triplets. The first number of each triplet is decreased by 1: −12 6 4 −13 7 3 −14 ___ 2. The second number of each triplet is increased by 1: −12 6 4 −13 7 3 −14 ___ 2. Thus, the missing number is 7 + 1 = 8. (Notice also that the 3rd number in each triplet is decreased by 1: −12 6 4 −13 7 3 −14 ___ 2.)

18. c. This is an arithmetic series in which each number is increased by 125. The missing number will be 5,798 + 125, or 5,923.

19. d. The pattern here is +5, +5, repeat, +5, +5, repeat. See below:
6 (+5) 11 (+5) 16 (repeat ➔) 16 (+5) 21 (+5) 26 (repeat ➔) 26 (+5) ___. Thus, the missing number is 26 + 5 = 31.

20. c. The pattern here is + 4, × 6, + 4, × 6, and so forth. See below:
10 (+ 4) 14 (× 6) 84 (+ 4) 88 (× 6) 264 (+ 4) ___
Thus, the missing number is 264 + 4 = 268.

21. b. Here, the numbers are decreasing, though not by a steady amount or by a common ratio. The pattern of decrease is: 38 (minus 3 × 6) 20 (minus 3 × 5) 5 (minus 3 × 4) −7 (minus 3 × 3) −16 (minus 3 × 2) ___
Thus, the missing number is −16 minus 3 × 2, or −16 − 6 = −22.

22. a. Here, the pattern is −1, × 2, − 1, × 2, and so forth:
9 (− 1) 8 (× 2) 16 (− 1) 15 (× 2) ___ (− 1) 29 (× 2) 58
Thus, the missing number is 15 × 2 = 30. You can check that you are right by subtracting 30 − 1 = 29, which is the next number in the series.

23. c. In this series, two numbers are repeated, then 13 is subtracted to arrive at the next number. Thus, the missing number is 53 − 13 = 40.

24. c. This is a multiplication series with repetition. The decimal (0.2, 0.4, 0.8) is repeated by a fraction with the same value ($\frac{1}{5}$, $\frac{2}{5}$, $\frac{4}{5}$) and is then multiplied by 2. Thus, the next number will be .8 × 2, or 1.6.

25. c. In this simple arithmetic series, each number increases by 0.8. Thus, the next number should be 3.9 + .8 = 4.7, choice **c**.

26. **c.** In this simple alternating addition and subtraction series, 2 is subtracted, then 1 is added, and so on. Thus, the next number should be 25 + 1, or 26.

27. **a.** This is a simple alternating subtraction series, which subtracts 2, then 5. Thus, the next number will be 17 – 2 = 15.

28. **a.** This is a simple alternating addition and subtraction series. The first series begins with 10 and adds 2 (10, 12, 14, 16); the second begins with 34 and subtracts 3 (34, 31, 28). Thus, the number that belongs in the blank is 14.

29. **a.** Each number in the pattern is one-half of the previous number. Half of 15 is $7\frac{1}{2}$. You can check the pattern by taking half of $7\frac{1}{2}$, which is $3\frac{3}{4}$, the next term.

30. **d.** This alternating addition series begins with 3. 1 is added to give 4; then 3 is added to give 7; then 1 is added, and so on. Thus, the next number will be 12 + 3 = 15.

31. **a.** This is an alternating series. In the first pattern, 8 is added (1, 9, 17); in the second pattern, 1 is added (4, 5, 6). Thus, the next number will be 6.

32. **b.** This is a simple subtraction series. Each number decreases by $\frac{1}{8}$. The next number is $\frac{5}{8} - \frac{1}{8}$, which is $\frac{4}{8}$, or $\frac{1}{2}$.

33. **c.** This is an alternating repetition series, with a random number, 22, introduced as every third number into an otherwise simple addition series. In the addition series, 4 is added to each number to arrive at the next number. Thus, the next two numbers will be 22 (the random number) followed 24 + 4, or 28.

34. **d.** Given the pattern, $\frac{1}{2}, \frac{1}{4}, \frac{1}{8}, \frac{1}{16} \ldots$ notice that the denominators double as the pattern advances. There are 4 terms so far. The fifth term will have a denominator of 32, the sixth term will be 64, the seventh term will be 128, the eighth term will be 256, the ninth term will be 512, and the tenth term will be 1,024. So the tenth term is $\frac{1}{1{,}024}$.

35. **b.** This is an alternating multiplication and subtraction series: First, multiply by 2, and then subtract 8. The next term will be 64 – 8 = 56.

36. **d.** This is a simple alternating addition and subtraction series. First, 3 is added, then 1 is subtracted; then 3 is added, 1 subtracted, and so on. Thus, the next term will be 15 + 3 =18. The term after that will be 18 – 1 = 17.

37. **a.** This is a simple addition series with a random number, 21, introduced as every third number. In the series, 6 is added to each number except 21, to arrive at the next number. The next number is the random number 21.

38. **a.** This is a simple subtraction series; each number is 4 less than the previous number. XX = 20, XVI = 16, XII = 12, VIII = 8, so the next number should be 4. In Roman numerals, 4 is written as IV, choice **a**.

39. **c.** In this series, the letters progress by 2 (J, L, N, P), while the numbers decrease by 3 (14, 11, 8, 5). Thus, the next term will be R2, choice **c**.

40. **b.** This is an alternating addition and subtraction series. Roman numbers alternate with Arabic numbers. In the Roman numeral pattern, each number decreases by 1 (VI, V, IV, III, corresponding to 6, 5, 4, 3) . In the Arabic numeral pattern, each number increases by 1 (10, 11, 12, 13). Thus, the next number should be the Roman numeral for 3, which is III.

41. **b.** If you consider each triplet of letters, the first letter in each triplet progresses from J→K→L→M→____. The second letter in each triplet progresses from A→B→C→D→____, and the third letter in each triplet progresses from K→L→M→N→____. Therefore, the last triplet should be NEO.

42. **a.** If you look carefully at this sequence, you will notice that the *entire* sequence is the alphabet (starting at C) written *backwards*. Therefore, the missing three letters are HGF.

43. **d.** If you look at the first letter in each quadruplet, you can see that one letter is skipped: ELFA GLHA ILJA _____ MLNA, so the first missing letter is K. Looking at the second letter in each quadruplet, you see that the letter L is constant: ELFA GLHA ILJA _____ MLNA, so the second missing letter must be L. Next, look at the third letter in each quadruplet: ELFA GLHA ILJA _____ MLNA. Again, one letter is skipped, so the missing letter is L. Finally, look at the last letter in each quadruplet: ELFA GLHA ILJA _____ MLNA. The letter A is a constant, so the last missing letter is A. Thus, the entire missing piece is KLLA.

44. **b.** Notice that each group of symbols has three versions of the same shape, the middle version being the largest: ○◯● | ▲△▵ | □ ___. Also, a black and white version of the shape borders this large middle shape. Notice that the circle is on the right and the black triangle is on the left. The missing shapes will be squares (thus choice **c** is wrong). The next two shapes will be a large square with the black square on the right: □■.

45. **a.** The first group contains a *square* between 2 triangles. Next, there is a circle between 2 *squares*. Third, there is a ***diamond*** surrounded by 2 circles. The last set has a rectangle in the middle. It should be surrounded by 2 ***diamonds***.

46. **b.** This is simply an alternating pattern. First, the 2 arrows point right, then one points up and one points down. Thus, the next part of the sequence should contain the 2 arrows pointing right.

47. **d.** This is a symbol series question. The first image is reflected (flipped), generating the second image. Then, the second is flipped to form the third. Thus, the fourth image will be the reflection of l⌐ which will look like this: ⌐l.

48. **a.** Look at the number of dots on each domino in each triplet: ⚄ ⚂ ⚀ | ⚀ ⚂ ⚄ | ___ ___ ⚀. The first triplet has 5 dots, 3 dots, 1 dot. The next triplet has 1 dot, 3 dots, 5 dots. The last triplet ends with 1 dot. It is safe to assume that the pattern here is 1-3-5; 5-3-1; 1-3-5, so the 2 missing dominos are ⚄ ⚂, the 5 and the 3.

49. **c.** Notice that the first and the third segments are upside-down versions of each other. The second and the fourth should also be upside-down versions of each other. Thus, the missing piece of the last segment looks like this: ш.

50. **c.** The first and the third sets of figures are inversions. They swap the inner shape for the outer shape. The second and fourth would then be expected to swap the top and bottom shapes. Thus, we would expect the missing shape to be a square on top of a circle, choice **c**.

Percents

WHAT IS A PERCENT?

Percents are a way of expressing values out of 100. For example 30% (30 percent) is equivalent to 30 out of 100 or $\frac{30}{100}$. Thus, you can express a percent as a fraction by placing the value before the percent symbol over 100. You can express a percent as a decimal by moving the current decimal point 2 places to the left. For example, 30% is also equivalent to .30.

You can convert a decimal value into an equivalent percent by moving the current decimal point 2 places to the right. For example, .30 = 30 %. This makes sense because percents are just *hundredths*, so .30 is 30 *hundredths*, or $\frac{30}{100}$, otherwise known as 30%.

Fractions can be converted to percentages by converting to a denominator of 100. This can be done by setting up a simple proportion. For example, to convert $\frac{2}{5}$ into an equivalent percentage, we set up this proportion:

$$\frac{2}{5} = \frac{?}{100}$$

Cross-multiply to get $2 \times 100 = 5 \times ?$, or $200 = 5 \times ?$. Divide both sides by 5 to get $? = 40$. Thus, $\frac{2}{5}$ is equivalent to 40%.

Sample Question:

$\frac{17}{20}$ is equivalent to what percent?

a. 17%
b. 65%
c. 85%
d. 90%

First, set up a proportion:

$$\frac{17}{20} = \frac{?}{100}$$

Cross-multiply to get $17 \times 100 = 5 \times ?$, or $1{,}700 = 20 \times ?$. Divide both sides by 20 to get $? = 85$. Thus, the answer is 85%, choice **c.**

▶ TAKING THE PERCENT OF A NUMBER

When you are calculating the percent of a number, just remember that *of* means multiply. Thus 50% of 40 is $50\% \times 40$. You can convert 50% to .50 and multiply $.50 \times 40 = 20$.

To save time, you should be familiar with the following equivalencies:

Fraction	Percent
$\frac{1}{5}$	20%
$\frac{1}{4}$	25%
$\frac{1}{3}$	Approximately 33%
$\frac{1}{2}$	50%
$\frac{2}{3}$	Approximately 66%
$\frac{3}{4}$	75%

Sample Question:

What is 75% of 400?

a. 300
b. 275
c. 100
d. 30

Remember that "of" means multiply. 75% of 400 is equivalent to 75% × 400. Because 75% = $\frac{3}{4}$, you write $\frac{3}{4}$ × 400. This equals 300, choice **a**.

UNKNOWN PERCENTS

When you do not know the percent of a value, you can express this percent as $\frac{?}{100}$. This means that when you see the phrase *what percent* you can express this mathematically as $\frac{?}{100}$.

Sample Question:

What percent of 800 is 40?

a. .05%
b. 5%
c. 15%
d. 50%

What percent is expressed mathematically as $\frac{?}{100}$. *Of* means *multiply*. *Is* means *equals*. Thus, the question, "What percent of 800 is 40?" can be rewritten as $\frac{?}{100 \times 800} = 40$. Solving, you get $\frac{? \times 800}{100} = 40$; $? \times 8 = 40$; $? = 5$. Thus, the answer is 5%, choice **b**.

PERCENT CHANGE, PERCENT ERROR, AND PERCENT PROFIT OR LOSS

When calculating a **percent change** (such as a **percent increase or decrease)** you simply express the ratio of the change to the initial as a value over 100. The general proportion to use is:

$$\frac{\text{Change}}{\text{Initial}} = \frac{?}{100}$$

Similarly, when calculating the **percent error**, you set a proportion that equates the difference between the calculated value and the actual value to the actual value with an unknown out of 100:

$$\frac{\text{Difference in values}}{\text{Actual value}} = \frac{?}{100}$$

When setting up a proportion to calculate percent profit or loss, you create a ratio of the net profit (or loss) to the initial cost and set this ratio equal to an unknown out of 100:

$$\frac{\text{net profit}}{\text{initial}} = \frac{?}{100} \qquad \frac{\text{net loss}}{\text{initial}} = \frac{?}{100}$$

Sample Question:

A business spent $1,000 on a shipment of products. The products were sold for only $750—a loss for the company. What is the percent loss?

a. 50%

b. 25%

c. 20%

d. 15%

Use the proportion:

$$\frac{\text{net loss}}{\text{initial}} = \frac{?}{100}$$

The net loss is 1,000 – 750 = 250 dollars and the initial amount was 1,000. The proportion becomes: $\frac{250}{1{,}000} = \frac{?}{100}$. Cross-multiplying yields 250 × 100 = 1,000 × *?*, or 25,000 = 1,000 × *?*, and *?* = 25. Thus, the answer is **b.**

SIMPLE AND COMPOUND INTEREST

The formula for simple interest is:

$$I = PRT$$

The amount of money deposited is called the principal, *P*. The interest rate per year is represented by *R*, and *T* represents the time in years.

When calculating compound interest, it is easiest to sequentially calculate the interest earned using *I* = *PRT*. You should be familiar with the following ways of compounding interest:

- **Compounded annually**: interest is paid each year
- **Compounded semi-annually**: interest is paid two times per year
- **Compounded quarterly:** interest is paid four times a year
- **Compounded monthly**: interest is paid every month
- **Compounded daily:** interest is paid every day

Sample Question:

If Howard puts $30,000 in the bank at a 4% rate of interest per year, how much interest will he make in 6 months?

a. $400

b. $600

c. $720

d. $7,200

The correct answer is choice **b.** Use the formula $I = PRT$. Where P equals \$30,000, $R = 4\% = .04$, and $T = \frac{1}{2}$ a year. Note that you must convert the 6 months into years. The formula becomes: $I = PRT = 30{,}000 \times .04 \times \frac{1}{2} = \$600$.

PRACTICE QUESTIONS

1. 15% is equivalent to which fraction below?

a. $\frac{3}{20}$

b. $\frac{15}{1{,}000}$

c. $\frac{1}{5}$

d. $\frac{1}{15}$

2. 20% is equivalent to which decimal value below?

a. .020

b. 2.0

c. 0.2

d. .002

3. When converted to a decimal, 45% is equivalent to

a. .045

b. .45

c. 4.5

d. 45

4. 73% can be expressed as which of the following fractions?

a. $\frac{.73}{100}$

b. $\frac{73}{100}$

c. $\frac{73}{1{,}000}$

d. $\frac{.73}{.10}$

5. 1.5% is equivalent to which decimal value below?

a. .15

b. 1.5

c. .0015

d. .015

6. When expressed as a percent, $\frac{31}{50}$ is equivalent to

a. 62%

b. $\frac{31}{50}$%

c. $\frac{3}{5}$%

d. 31%

7. Another way to write 26.5% is

a. $\frac{.265}{100}$

b. $\frac{26}{80}$

c. $\frac{53}{200}$

d. $\frac{26.5}{1000}$

8. .0037% is equivalent to which of the following fractions?

a. $\frac{37}{1{,}000}$

b. $\frac{37}{10{,}000}$

c. $\frac{37}{1{,}000{,}000}$

d. $\frac{37}{10{,}000{,}000}$

9. Which of the following is 17% of 6,800?

a. 200

b. 340

c. 578

d. 1,156

10. Which number sentence below is false?

a. $20\% \leq \frac{1}{5}$

b. $25\% = \frac{2}{8}$

c. $35\% > \frac{24}{50}$

d. $\frac{3}{4} \leq 80\%$

11. Express 12 out of 52 to the nearest percent.

a. 23%

b. 24%

c. 25%

d. 26%

12. $\frac{4}{5}$% is equal to
a. 80
b. 8
c. .08
d. .008

13. 50% of what number equals 20% of 2000?
a. 200
b. 400
c. 600
d. 800

14. 300% of 54.2 equals
a. 16.26
b. 162.6
c. 1,626
d. none of the above

15. What percent of $\frac{1}{2}$ is $\frac{1}{8}$?
a. 25%
b. 50%
c. 80%
d. none of the above

16. To calculate 75% of a dollar amount, you can
a. multiply the amount by 75
b. divide the amount by 75
c. multiply the amount by $\frac{3}{4}$
d. divide the amount by $\frac{3}{4}$

17. 40% of what number is equal to 460?
a. 575
b. 640
c. 860
d. 1,150

18. Larry makes a 12% commission on every car he sells. If he sold $40,000 worth of cars over the course of three months, what was his commission on these sales?
a. $44,800
b. $35,200
c. $8,000
d. $4,800

19. Zip drives cost $100 each. When more than 50 are purchased, an 8% discount is applied. At a store that charges 8% tax, how much money will 62 zip drives cost? (Round to the nearest cent.)

a. $6,200
b. $6,160.32
c. $5,704
d. $456.32

20. Mike made $64,000 in 2002, but he had to pay 26% tax on that amount. How much did he make after taxes?

a. $80,640
b. $67,640
c. $47,360
d. $42,360

21. What percent of $\frac{8}{9}$ is $\frac{2}{3}$?

a. 33%
b. 66%
c. 75%
d. 80%

22. 400 books went on sale this week. So far, 120 were sold. What percent of the books remain?

a. 15%
b. 30%
c. 70%
d. 80%

23. What percent of the circle below is shaded?

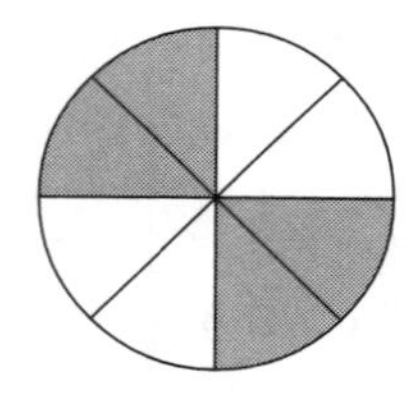

a. 25%
b. 50%
c. 75%
d. 100%

24. What percent of the square below is shaded?

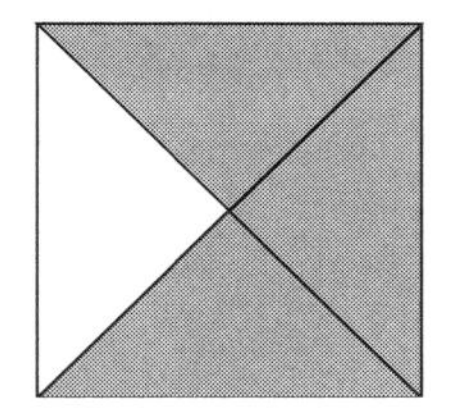

a. 25%
b. 50%
c. 75%
d. 100%

25. What percent of the square below is shaded?

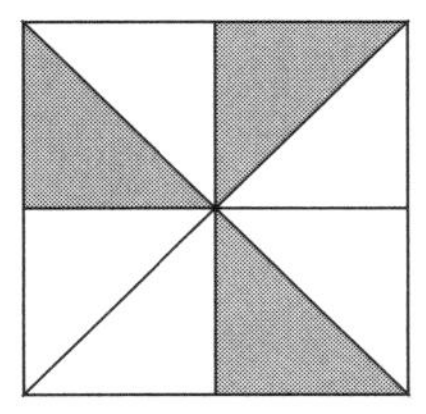

a. 20%
b. 37.5%
c. 40%
d. 80%

26. What percent of the square below is shaded?

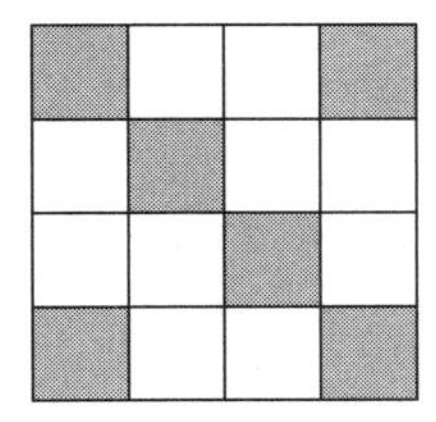

a. 20%
b. 37.5%
c. 40%
d. 80%

27. A dealer buys a car from the manufacturer for $13,000. If the dealer wants to earn a profit of 20% based on the cost, at what price should he sell the car?

a. $16,250
b. $15,600
c. $15,200
d. $10,833

28. 33 is 12% of which of the following?

a. 3,960
b. 396
c. 275
d. 2,750

29. Of the numbers listed, which choice is not equivalent to the others?

a. 52%
b. $\frac{13}{25}$
c. 52×10^{-2}
d. .052

30. Use the formula $I = PRT$ to answer the following question:
Gary Otto made $8,000 and put half that amount into an account that earned interest at a rate of 6% per year. After 2 years, what is the dollar amount of the interest earned?

a. $4,800
b. $960
c. $660
d. $480

31. If Kamil puts $10,000 in the bank at a 6% rate of interest per year, how much interest will he make in 8 months? Use the formula $I = PRT$.

a. $400
b. $350
c. $300
d. $250

32. If Veronica deposits $5,000 in an account with a yearly interest rate of 9%, and leaves the money in the account for 8 years, how much interest will her money earn?

a. $360,000
b. $45,000
c. $3,600
d. $450

33. At the city park, 32% of the trees are oaks. If there are 400 trees in the park, how many trees are NOT oaks?

a. 128
b. 272
c. 278
d. 312

34. Which ratio best expresses the following: five hours is what percent of a day?

a. $\frac{5}{200} = \frac{x}{24}$

b. $\frac{5}{24} = \frac{24}{x}$

c. $\frac{5}{24} = \frac{x}{100}$

d. $\frac{x}{100} = \frac{24}{5}$

35. If 10% of a number is 45, what would 20% of that number be?

a. 9

b. 90

c. 450

d. 900

36. A dozen staplers cost $10.00 and will be sold for $2.50 each. What is the rate of profit?

a. 75%

b. 100%

c. 150%

d. 200%

37. A statue was bought at a price of $50 and sold for $38. What is the percent loss?

a. 12%

b. 15%

c. 24%

d. 30%

38. The price of a $130 jacket was reduced by 10% and again by 15%. What is the new cost of the jacket?

a. $97.50

b. $99.45

c. $117

d. $125

39. At an electronics store, all items are sold at 15% above cost. If the store purchased a printer for $85, how much will they sell it for?

a. $90

b. $98.50

c. $97.75

d. $95.50

40. Mark paid $14,105 for his new car. This price included 8.5% for tax. What was the price of the car excluding tax?

a. $13,000
b. $13,850
c. $11,989.25
d. $1,198.93

41. Steven's income was $34,000 last year. He must pay $2,380 for income taxes. What is the rate of taxation?

a. 8%
b. 7%
c. .008%
d. .007%

42. $8,000 is deposited into an account. If interest is compounded semiannually at 5% for 1 year, how much money is in the account at the end of the year?

a. $8,175
b. $8,200
c. $8,400
d. $8,405

43. $14,000 is deposited into an account. If interest is compounded quarterly at 8% for 9 months, how much money will be in the account at the end of this period?

a. $14,280.00
b. $14,565.60
c. $14,856.91
d. $15,154.05

44. Sam has $1,000 to invest. He would like to invest $\frac{3}{5}$ of it at 6% simple interest. The remainder would be invested at 8% simple interest. How much interest would he have earned after one year?

a. $32
b. $36
c. $68
d. $70

45. How many twelfths are there in $33\frac{1}{3}\%$?

a. 1
b. 4
c. 33
d. 100

46. What is the percent increase from 150 to 200?

a. 25%
b. $33\frac{1}{3}\%$
c. 75%
d. $66\frac{2}{3}\%$

47. What is the percent decrease from 200 to 150?

a. 25%
b. $33\frac{1}{3}\%$
c. 75%
d. $66\frac{2}{3}\%$

48. If a crate weighing 600 pounds weighs 540 pounds on a broken scale, what is the percent error?

a. 10%
b. 11%
c. 15%
d. 25%

49. A five-gallon tank is completely filled with a solution of 50% water and 50% alcohol. Half of the tank is drained and 2 gallons of water are added. How much water is in the resulting mixture?

a. 2.5 gallons
b. 3.25 gallons
c. 3.5 gallons
d. 4.5 gallons

50. Steve earned a $4\frac{3}{4}\%$ pay raise. If his salary was $27,400 before the raise, how much was his salary after the raise?

a. $27,530.15
b. $28,601.50
c. $28,701.50
d. $29,610.50

ANSWERS

1. **a.** 15% equals $\frac{15}{100}$. $\frac{15}{100}$ reduces to $\frac{3}{20}$.

2. **c.** To change 20% to its equivalent decimal form, move the decimal point two places to the left. Thus, 20% = .20. Choice **c**, 0.2 is equivalent to .20.

3. **b.** When you see a percent symbol (%), you just move the decimal point 2 places to the left. Thus, 45% is equivalent to .45.

4. **b.** When you see a percent symbol (%), you can rewrite the percent as a fraction by placing the value over 100. Thus, 73% is equivalent to $\frac{73}{100}$.

5. **d.** 1.5% can be converted to its equivalent decimal form by moving its decimal point 2 places to the left. Thus, 1.5% is equivalent to .015, choice **d**.

6. **a.** When written as fractions, percents have a denominator of 100. You can easily convert $\frac{31}{50}$ to a fraction with a denominator of 100 by multiplying by $\frac{2}{2}$. $\frac{31}{50} \times \frac{2}{2} = \frac{62}{100}$ = 62%, choice **a**.

7. **c.** First, put 26.5 over 100 = $\frac{26.5}{100}$. This is not an answer choice, so we need to reduce. Multiply $\frac{26.5}{100}$ by $\frac{10}{10}$ before reducing: $\frac{26.5}{100} \times \frac{10}{10} = \frac{265}{1000}$. Now, we reduce $\frac{265}{1,000} = \frac{53}{200}$.

8. **c.** To change a percent to a fraction, just place the value before the percent symbol of 100. Thus, .0037% = $\frac{.0037}{100}$. In order to get a whole number in the numerator, multiply the fraction by $\frac{10,000}{10,000}$. Thus, $\frac{.0037}{100} \times \frac{10,000}{10,000} = \frac{37}{1,000,000}$.

9. **d.** We need to find 17%, or .17 of 6,800. Remember that *of* means multiply: .17 × 6,800 = 1,156.

10. **c.** 20% = $\frac{20}{100}$, or $\frac{1}{5}$, so choice **a** represents a true statement. 25% = $\frac{25}{100} = \frac{1}{4}$, and $\frac{2}{8} = \frac{1}{4}$, so choice **b** is also true. In choice **c**, 35% = $\frac{35}{100}$ and $\frac{24}{50} = \frac{48}{100}$. Thus, the statement 35% > $\frac{24}{50}$ is not true. Choice **c** is, therefore, the correct answer. In choice **d**, $\frac{3}{4}$ = 75%, which is in fact less than 80%.

11. **a.** "12 out of 52" is written as $\frac{12}{52}$. Set up a proportion to see how many hundredths $\frac{12}{52}$ is equivalent to: $\frac{12}{52} = \frac{?}{100}$. Cross-multiplying yields 100 × 12 = 52 × *?*, or 1,200 = 52 × *?*. Dividing both sides by 52 yields *?* = 23.07623. When expressed to the nearest percent, this rounds to 23%.

12. **d.** It is easier to change $\frac{4}{5}$ into .8 before dealing with the percent symbol. $\frac{4}{5}$% = .8% = .008.

13. **d.** "50% of what number equals 20% of 2,000?" can be written mathematically as .50 × *?* = .20 × 2,000. Dividing both sides by .5 will yield *?* = $\frac{(.2)(2,000)}{5}$ = 800.

14. **b.** 300% equals $\frac{300}{100}$, or 3. To find 300% of 54.2, just multiply 3 times 54.2: 3 × 54.2 = 162.6.

15. **a.** "What percent" can be expressed as $\frac{?}{100}$. The question, "What percent of $\frac{1}{2}$ is $\frac{1}{8}$?" can be expressed as: $\frac{?}{100} \cdot \frac{1}{2} = \frac{1}{8}$. This simplifies to $\frac{?}{200} = \frac{1}{8}$; cross-multiplying yields 8 × *?* = 200; dividing both sides by 8 yields 25.

16. **c.** 75% = $\frac{75}{100}$. This reduces to $\frac{3}{4}$. Taking $\frac{3}{4}$ of a dollar amount means you multiply the dollar amount by $\frac{3}{4}$.

17. **d.** The question, "40% of what number is equal to 460?" can be written mathematically as: .40 × *?* = 460. Next, divide both sides by .40 to yield *?* = 1,150.

18. **d.** He gets 12% of $40,000, or .12 × $40,000 = $4,800.

19. **b.** Since more than 50 drives are being purchased, use the discounted price. Take 8% ($8) off the cost of each drive. So, instead of costing $100 each, the drives will be $92 each. Next, multiply 62 drives by the price of each drive: 62 × 92 = $5,704. Next, calculate the tax. $5,704 × .08 = $456.32. Add the tax to the $5,704 to get $6,160.32.

20. **c.** The tax on the $64,000 will equal .26 × 64,000 = $16,640. Subtract the tax from his earnings: 64,000 – 16,640 = $47,360.

21. **c.** The question, "What percent of $\frac{8}{9}$ is $\frac{2}{3}$?" can be expressed mathematically as $\frac{?}{100} \times \frac{8}{9} = \frac{2}{3}$. Divide both sides by $\frac{8}{9}$ to get $\frac{?}{100} = \frac{2}{3} \div \frac{8}{9}$ or $\frac{?}{100} = \frac{2}{3} \times \frac{9}{8}$. This simplifies to $\frac{?}{100} = \frac{18}{24}$, or $\frac{?}{100} = \frac{3}{4}$. Multiply both sides by 100 to get $? = \frac{300}{4}$, so $? = 75$.

22. **c.** 120 out of a total of 400 were sold. Simply set up a proportion to see what this would be equivalent to when expressed out of 100.

$$\frac{120}{400} = \frac{?}{100}$$

Cross-multiplying, we get $120 \cdot 100 = 400 \times ?$, which is the same as $12{,}000 = 400 \times ?$, and dividing both sides by 400 yields $? = 30$. Thus, 30% were sold, so 70% remain.

23. **b.** $\frac{1}{2}$ of the circle is shaded. $\frac{1}{2} = \frac{50}{100} = 50\%$.

24. **c.** $\frac{3}{4}$ of the square is shaded. $\frac{3}{4} = \frac{75}{100} = 75\%$.

25. **b.** $\frac{3}{8}$ of the square is shaded. 3 ÷ 8 = .375. To express this as a percent, move the decimal two places to the right: 37.5%.

26. **b.** $\frac{6}{16}$ of the square is shaded. $\frac{6}{16}$ reduces to $\frac{3}{8}$. 3 ÷ 8 = .375. To express this as a percent, move the decimal two places to the right: 37.5%.

27. **b.** A 20% mark-up yields a new price that is 120% of the original price. $13,000 × 1.20 = $15,600.

28. **c.** "33 is 12% of what number?" can be expressed mathematically as $33 = .12 \times ?$. Just divide 33 by 0.12 (12 percent) to get 275.

29. **d.** 52% is the same as .52 (drop the % sign and move the decimal point two places to the left). $\frac{13}{25} = \frac{26}{50} = \frac{52}{100}$. 52 ÷ 100 = .52. And $52 \times 10^{-2} = 52 \times .01 = .52$. Obviously, .052 does not equal .52, so your answer is **d.**

30. **d.** $I = PRT$ means *Interest = principal × rate of interest × time*. Principal = your original amount of money (in dollars), and time is in years. Be careful, the original amount of money (P) is $4,000 because Gary put $\frac{1}{2}$ of the $8,000 into the account. $I = .06$ and $T = 2$ years. Substituting into $I = PRT$, you get $I = (4{,}000)(.06)(2) = \$480$.

31. **a.** The formula $I = PRT$ means: *Interest = principal × rate of interest × time* (Where principal = your original amount of money (in dollars), and time is in years.) Here, we were given the time frame of 8 months, so we need to convert to years. 8 months × $\frac{1 \text{ yr}}{12 \text{ months}} = \frac{8}{12}$ yr = $\frac{2}{3}$ yr. We are given P = \$10,000 and R = 6% or .06. Next, we substitute these values into the equation:

$I = PRT$

$I = (\$10{,}000)(.06)(\frac{2}{3})$

$= 600 \times \frac{2}{3}$

$= \frac{1{,}200}{3}$

$= \$400$.

32. **c.** In the formula $I = PRT$, the amount of money deposited is called the principal, P. The interest rate per year is represented by R, and T represents the number of years. The interest rate must be written as a decimal. Here $P = 5{,}000$, $R = 9\% = .09$, and $T = 8$. Substitute these numbers for the respective variables and multiply: $I = 5{,}000 \times .09 \times 8 = \$3{,}600$.

33. **b.** First, determine what percent of the trees are not oaks by subtracting. 100% – 32% = 68%. Change 68% to a decimal (.68) and multiply: 0.68 × 400 = 272.

34. **c.** The problem can be restated as: 5 hours is to 24 hours as x% is to 100%. This is the same as: $\frac{5}{24} = \frac{x}{100}$.

35. **b.** First figure out what the number is. If 10% of a number is 45, we can call the number "*?*" and write .10 × *?* = 45. Divide both sides by .10 to get *?* = 450. Next, take 20% of 450: .20 × 450 = 90.

36. **d.** When all of the staplers sold, the amount collected is \$2.50 × 12 = \$30. Since a dozen staplers cost \$10, the profit is \$20. Next, set up a proportion:

$$\frac{\$20 \text{ profit}}{\text{initial } \$10} = \frac{?}{100}$$

Cross-multiply to get (100)(20) = (10)(*?*), or 2,000 = (10)(*?*). Divide both sides by 10 to get *?* = 200. Thus, the rate of profit is 200%.

37. **c.** Find the net loss: \$50 – \$38 = \$12. Next, set up a proportion:

$$\frac{\$12 \text{ loss}}{\text{initial } \$50} = \frac{?}{100}$$

Cross-multiply to get 12 × 100 = 50 × *?*, or 1,200 = 50 × *?*. Divide both sides by 50 to get *?* = 24. Thus, there is a 24% loss.

38. **b.** \$130 – 10% of 130 = 130 – 13 = \$117. Next, take 15% of \$117 = .15 × 117 = \$17.55. Deduct this amount: \$117 – \$17.55 = \$99.45. Choice **a**, \$97.50, is wrong because this represents a 25% reduction in price. You cannot add 10% and 15%, and deduct 25%.

39. **c.** The printer will sell for 115% of the cost. 115% × \$85 = 1.15 × 85 = 97.75. This question can also be solved in two steps: 15% of 85 = \$12.75 markup. Add \$12.75 to \$85 (the cost) to get \$97.75.

40. **a.** If the price of the car is p, then you know that the price of the car plus 8.5% of that price added up to \$14,105. 8.5% equals .085. Thus, $p + .085p = 14{,}105$. $1.085p = 14{,}105$. Dividing both sides by 1.085 yields $p = \$13{,}000$.

41. b. You can solve this problem by asking yourself, "2,380 is what percent of 34,000?" and then expressing this question mathematically: $2{,}380 = \frac{?}{100} \times 34{,}000$. Divide both sides by 34,000 to get $\frac{2{,}380}{34{,}000} = \frac{?}{100}$. Cross-multiply to get 238,000 = (34,000)(*?*). Divide both sides by 34,000 to get 7. Thus, the answer is 7%.

42. d. Because the interest is compounded semiannually (twice a year), after $\frac{1}{2}$ a year the amount of interest earned $I = PRT = 8{,}000 \times .05 \times \frac{1}{2} = \$200$. Now the account has $8,200 in it. Next, calculate the interest for the second half of the year with $I = PRT = 8{,}200 \times .05 \times \frac{1}{2} = 205$. Thus, the answer is $8,405.

43. c. Note that 9 months = $\frac{3}{4}$ of a year. Because interest is compounded quarterly (4 times a year), after $\frac{1}{4}$ of a year, the amount of interest earned will be $I = PRT = 14{,}000 \times .08 \times \frac{1}{4} = \$280$. The amount in the account after this time will be $14,280. After another $\frac{1}{4}$ of a year, we add $I = PRT = 14{,}280 \times .08 \times \frac{1}{4} = \$285.60$. The new total is $14,565.60. After the next $\frac{1}{4}$ of a year, the amount of interest earned is $I = PRT = 14{,}565.60 \times .08 \times \frac{1}{4} = \$291.312$. The amount in the account after $\frac{3}{4}$ of a year is $14,856.91.

44. c. Because Sam is making 2 investments, first find $\frac{3}{5}$ of $1,000. Divide $1,000 into 5 equal parts ($\frac{\$1{,}000}{5} = \$200$) and take 3 parts ($600). $600 is invested at 6% simple interest, which yields:
$600 (6%) = $600 (.06) = $36.
The remaining $400 is invested at 8% simple interest, which yields:
$400 (8%) = $400 (.08) = $32.
The total interest earned is $36 + $32 = $68.

45. b. Convert $33\frac{1}{3}\%$ into a fraction, remembering that the percent sign is equivalent to $\frac{1}{100}$. $33\frac{1}{3}\% \times \frac{1}{100} = \frac{1}{3}$. Now, $\frac{1}{3} = \frac{4}{12}$. Therefore, there are 4 twelfths in $33\frac{1}{3}\%$.

46. b. Use the proportion:

$$\frac{\text{Change}}{\text{Initial}} = \frac{?}{100}$$

Where the change = 200 – 150 = 50, and the initial value is 150. Thus, we have:

$$\frac{50}{150} = \frac{?}{100}$$

Cross-multiply to get 50 × 100 = 150 × *?*, or 5,000 = 150 × *?*. Divide both sides by 150 to get $? = 33\frac{1}{3}$. Thus, there was a $33\frac{1}{3}\%$ increase.

47. a. Use the proportion:

$$\frac{\text{Change}}{\text{Initial}} = \frac{?}{100}$$

Where the change = 200 – 150 = 50, and the initial value is 200. Thus, we have:

$$\frac{50}{200} = \frac{?}{100}$$

Cross-multiply to get 50 × 100 = 200 × *?*, or 5,000 = 200 × *?*. Divide both sides by 200 to get *?* = 25. Thus, there was a 25% decrease.

48. a. Use the proportion:

$$\frac{\text{Difference in values}}{\text{Actual value}} = \frac{?}{100}$$

Here, the difference in values is 600 lbs – 540 lbs = 60 lbs. The actual value is 600 lbs. Thus, we get:

$$\frac{60}{600} = \frac{?}{100}$$

Cross-multiplying yields 60 × 100 = 600 × *?*, or 6,000 = 600 × *?*. Divide both sides by 600 to get *?* = 10. Thus, there is a 10% error, choice **a**.

49. b. Draining half the 5-gallon tank leaves 2.5 gallons inside. Since you know the solution is a 50-50 mixture, there must be 1.25 gallons of water present at this point. After adding 2 gallons of water, there will be 1.25 + 2, or 3.25 gallons of water in the final mixture.

50. c. This problem requires both multiplication and addition. First, to determine the amount of the raise, change the percent to a decimal and multiply. 0.0475 × 27,400 = 1,301.5. Then, add this amount to the original salary. 1,301.50 + 27,400 = 28,701.50.

Word Problems

In addition to dealing with basic operations, fractions, decimals, and percents, common word problems on the Civil Service Exam involve distance, work and salaries, tank and pipe questions, labor questions, and ratio and proportions.

RATIOS AND PROPORTIONS

A *ratio* is a way of comparing two or more numbers. There are several different ways to write ratios. Here are some examples:

- with the word *to*: 1 to 2
- using a colon (:) to separate the numbers: 1:2
- using the term *for every*: 1 for every 2
- separated by a division sign or fraction bar: $\frac{1}{2}$

Usually, a fraction represents a part over a whole:

$\frac{part}{whole}$

Often, a ratio represents a part over a part:

$\frac{part}{part}$

But ratios can also represent a part over a whole:

$\frac{part}{whole}$

When a ratio represents a part over a part, you can often find the whole if you know all the parts. A ***proportion*** is a way of relating two ratios to one another. If you equate a given ratio to the part that you know, you can easily find an unknown part. Once you know the unknown parts, you can calculate the whole.

Sample Question:

If the ratio of union workers to non-union workers is 2:3 and there are 360 non-union workers, how many workers are there in all?

a. 240
b. 360
c. 600
d. 720

Here, you are given a 2:3 ratio. You know one part: that there are 360 non-union workers. You can set up a proportion in order to calculate the unknown part, $\frac{2}{3} = \frac{?}{360}$.

Cross-multiply to get $360 \cdot 2 = 3 \cdot ?$, or $720 = 3 \cdot ?$

Divide both sides by 3 to get $? = 240$. This is the missing part: the number of union workers. Finally, add the number of union workers to non-union workers to get the whole: $360 + 240 = 600$. Thus, the correct answer is **c**.

▶ WORK AND SALARIES

Some word problems deal with salaries. You should be familiar with the following salary schedules:

- **per hour**: amount earned each hour
- **daily**: amount earned each day
- **weekly**: amount earned each week
- **semi-weekly**: amount earned twice a week
- **semi-monthly**: amount earned twice a month
- **monthly**: amount earned each month
- **annually**: amount earned each year

Other problems involving work need to be dissected logically. For example, consider the following question.

Sample Question:

If 14 workers can complete a job in 2 days, how long will it take 4 workers to complete the same job? Assume all workers work at the same rate.

a. $\frac{4}{7}$ day
b. 5 days
c. 6 days
d. 7 days

Most people try to set up the following proportion when confronted with the above scenario:

$$\frac{14 \text{ workers}}{2 \text{ days}} = \frac{4 \text{ workers}}{? \text{ days}}$$

Notice that the *?* in the denominator of the second ratio will necessarily be smaller than the 2 days in the denominator of the first ratio. Does it make sense that 4 workers will be able to finish the job of 14 workers in less than 2 days?

This sort of question needs to be broken apart logically. If 14 workers can complete the job in 2 days, it will take one person 14 times as long to complete the same job: 28 days. It will take 4 people $\frac{1}{4}$ as long to complete this amount of work, or 7 days. Thus, choice **d** is correct.

TANK AND PIPE QUESTIONS

Tank and pipe questions must also be solved logically. Tank and pipe questions involve the filling and draining of tanks through various pipes. Once you see what the net (overall) effect is, you are able to solve the question posed to you.

Sample Question:

A tank is partly filled with water. Pipe X leads into the tank and can fill the entire tank in 4 minutes. Pipe Y drains the tank and can drain the entire tank in 3 minutes. At a certain point in time, the tank is halfway full, and the valves leading to pipes X and Y are closed. When these valves are opened simultaneously, how long will it take for the tank to drain?

a. 2 min.
b. 4 min.
c. 5 min.
d. 6 min.

First, consider Pipe X. It can fill the tank in 4 minutes. This means that for every minute that goes by, $\frac{1}{4}$ of the tank would get filled. Next, consider Pipe Y. This pipe can empty the tank in 3 minutes.

This means that for every minute that goes by, $\frac{1}{3}$ of the tank would get drained. When we consider these fractions as twelfths, we see that Pipe X fills $\frac{3}{12}$ per minute and Pipe Y drains $\frac{4}{12}$ per minute. The net effect is a draining of $\frac{1}{12}$ of the tank every minute. Since the tank starts out $\frac{1}{2}$ full (or $\frac{6}{12}$ full), it will take 6 minutes to drain the $\frac{6}{12}$ of water (at the rate of $\frac{1}{12}$ out per minute). Thus, choice **d** is correct.

DISTANCE

Distance questions can be solved with the formula $D = RT$, assuming that a constant rate is maintained. Here, you have the flexibility to use many different combinations of rates, distances, and times, so long as the units you use in the equation match each other. For example, rates can be measured in meters per second, kilometers per hour, feet per second, miles per hour, and so forth. Just be sure that if you use, for example, a rate in miles per hour as your R in the equation, that your D is in miles, and your T is in hours.

Sample Question:

Train A leaves its station and travels at a constant rate of 65 mph in an eastward direction. At the same time, Train B leaves a western station heading east at a constant rate of 70 miles an hour. If the two trains pass each other after 3 hours, how far apart were they initially?

a. 405 miles
b. 210 miles
c. 195 miles
d. none of the above

The correct answer is choice **c.** The 2 trains initial distance apart equals the sum of the distance each travels in 3 hours. Using $D = RT$, you know Train A travels a distance of (65)(3) = 195 miles, and Train B travels (70)(3) = 210 miles. This means that they were 195 + 210 = 405 miles apart initially. It is helpful to draw a diagram to understand this better:

Train A **Train B**

$D_A = RT$ $D_B = RT$

initial distance apart

PRACTICE QUESTIONS

Work and Salaries

1. Pete made $4,000 in January, $3,500 in February, and $4,500 in March. If he put 30% of his total earnings into his checking account and the rest into his saving account, how much money does he have in his checking account?

a. $3,600
b. $4,200
c. $6,300
d. $8,400

2. Denise had $120. She gave $\frac{1}{8}$ of this amount to Suzanne. She then gave $\frac{1}{4}$ of the remainder to Darlene. How much money does Denise have left?

a. $26.25
b. $30.00
c. $78.75
d. $80.00

3. Greg had $12,000 in his savings account. Of this amount, he transferred $\frac{1}{3}$ into checking, $\frac{1}{4}$ into a certificate of deposit, and spent $\frac{1}{8}$ on a computer system. How much money remains in his savings account?

a. $3,500
b. $5,000
c. $5,600
d. $6,000

4. If two pieces of wood measuring $2\frac{1}{2}$ feet and $3\frac{1}{3}$ feet are laid end to end, how long will their combined length be?

a. 5 feet 5 inches
b. 5 feet 10 inches
c. 6 feet 0 inches
d. 6 feet 5 inches

5. A shipment of cable weighs 3.2 lbs per foot. If the total weight of 3 identical reels of cable is 6,720 lbs, how many feet of cable are in each reel?

a. 64,512 feet
b. 21,504 feet
c. 2,000 feet
d. 700 feet

6. A school is purchasing 5 monitors at \$175 each, 3 printers at \$120 each, and 8 surge suppressors at \$18 each. If the school receives a 12% discount, what is the final cost (excluding tax)?

a. \$1,379.00
b. \$1,313.52
c. \$1,213.52
d. \$1,200.00

Ratios and Proportions

7. The Huntington Golf Club has a ratio of two women to every three men. A 2:3 ratio is equivalent to which of the following ratios?

a. 3:2
b. 4:8
c. 8:12
d. 4:12

8. A map drawn to scale shows that the distance between 2 towns is 3 inches. If the scale is such that 1 inch equals 1 km., how far away are the 2 towns in kilometers?

a. 3 miles
b. 3 km.
c. 30 miles
d. 30 km.

9. If it takes 27 nails to build 3 boxes, how many nails will it take to build 7 boxes?

a. 64
b. 72
c. 56
d. 63

10. Ralph can hike 1.3 miles in 45 minutes. Which equation could be used to find *d*, the distance in miles that Ralph can hike in 3 hours?

a. $\frac{d}{3} = \frac{0.75}{1.3}$
b. $\frac{1.3}{0.75} = \frac{d}{3}$
c. $\frac{0.75}{d} = \frac{3}{1.3}$
d. $\frac{0.75}{3} = \frac{d}{1.3}$

11. If Jack always spends \$18 on gaming equipment in a week, how much does he spend in 6 weeks?

a. \$60
b. \$48
c. \$108
d. \$180

12. If it takes a machine 5 minutes to build 3 components, how long would it take the same machine to build 18 components?

a. 90 min.
b. 18 min.
c. 15 min.
d. 30 min.

13. Dr. Martin sees an average of 2.5 patients per hour. If she takes an hour lunch break, about how many patients does she see during the typical 9 to 5 work day?

a. 16
b. 18
c. 20
d. 22

14. A diagram drawn to scale shows a diagonal of 12 cm. If the scale is 1.5 cm. = 1 foot, how long is the actual diagonal?

a. 8 ft.
b. 7.5 ft.
c. 6.8 ft.
d. 6 ft.

15. The height of the Statue of Liberty from foundation to torch is 305 feet 1 inch. Webster's American Mini-Golf has a 1:60 scale model of the statue. Approximately, how tall is the scale model?

a. 5 inches
b. 5 feet 1 inch
c. 6 feet 5 inches
d. 18,305 feet

Work and Salaries

16. Scott can pot 100 plants in 30 minutes. Henri can do the same job in 60 minutes. If they worked together, how many minutes would it take them to pot 200 plants?

a. 20 min.
b. 30 min.
c. 40 min.
d. 60 min.

17. Francine and Lydia are in the same book club, and both are reading the same 350-page novel. Francine has read $\frac{4}{5}$ of the novel. Lydia has read half as much as Francine. What is the ratio of the number of pages Lydia has read to the number of pages in the novel?

a. 1:2
b. 2:5
c. 2:3
d. 1:4

18. A construction job calls for $2\frac{5}{6}$ tons of sand. Four trucks, each filled with $\frac{3}{4}$ tons of sand, arrive on the job. Is there enough sand, or is there too much sand for the job?

a. There is not enough sand; $\frac{1}{6}$ ton more is needed.
b. There is not enough sand; $\frac{1}{3}$ ton more is needed.
c. There is $\frac{1}{3}$ ton more sand than is needed.
d. There is $\frac{1}{6}$ ton more sand than is needed.

19. Joseph earns a semi-monthly salary of $1,200. What is his yearly salary?

a. $144,000
b. $48,000
c. $28,800
d. $14,400

20. During a normal 40-hour work week, Mitch earns $800. His boss wants him to work this weekend and Mitch will get paid time and a half for these overtime hours. How much will Mitch make for 10 weekend hours?

a. $200
b. $240
c. $300
d. $340

21. Gary earns $22 an hour as a lab technician. Monday he worked 5 hours. Tuesday he worked 8 hours, and Wednesday he worked $4\frac{1}{2}$ hours. How much did he earn during those three days?

a. $363
b. $374
c. $385
d. $407

22. This month Ron earned $2,300 as his gross pay. Of this amount, $160.45 was deducted for FICA tax, $82.50 was deducted for state tax, $73.25 was deducted for city tax, and $100 was diverted to his 401K. How much was his net paycheck?

a. $1,883.80
b. $1,888.30
c. $1,983.80
d. $1,988.33

23. Two men can load a truck in 4 hours. How many trucks can they load in 6 hours?

a. 1
b. $1\frac{1}{2}$
c. 2
d. $2\frac{1}{2}$

24. A machine can assemble 400 parts in half an hour. Of the 400 parts, 5% will be defective. If 2 such machines are working, how many non-defective parts will be assembled in 5 hours?

a. 800
b. 1,600
c. 3,800
d. 7,600

25. Kate's daily salary is $120. If she worked 24 days this month, how much did she earn?

a. $3,600
b. $3,200
c. $3,000
d. $2,880

26. John earns $1,600 a month plus 8% commission on all sales. He sold $825 worth of merchandise during November, $980 worth of merchandise during December, and $600 work of merchandise during January. What was his total earning for these three months?

a. $1,792.40
b. $2,597.40
c. $1,924.00
d. $4,992.40

27. Four machines can complete a job in 6 hours. How long will it take 3 machines to complete the same job?

a. 4 hours
b. 8 hours
c. 10 hours
d. 12 hours

28. One construction job can be completed by 16 workers in 10 days. How many days would it take 8 workers to complete the job?

a. 12 days
b. 16 days
c. 18 days
d. 20 days

29. A job can be completed by 6 workers in 18 days. How many days would it take 9 workers to complete the job?

a. 12 days
b. 16 days
c. 18 days
d. 20 days

30. Nine workers working at the same pace can complete a job in 12 days. If this job must be completed in 3 days, how many workers should be assigned?

a. 27
b. 30
c. 36
d. 48

31. When Anthony and Dave work together they can complete a task in 3 hours. When Anthony works alone he can complete the same task in 8 hours. How long would it take for Dave to complete the task alone?

a. $6\frac{1}{2}$ hours
b. 6 hours
c. $4\frac{4}{5}$ hours
d. 4 hours

32. Rose and Marie worked on a project together. Rose put in 40 hours of work and Marie put in 60 hours of work. The contract for the entire project paid $2,000. The women decide to split the money according to the ratio of the amount of time each put into the project. How much did Marie get?

a. $400
b. $600
c. $1,000
d. $1,200

33. Al and Artie worked on a project together. Al put in 18 hours of work and Artie put in 24 hours of work. The contract for the entire project was $7,000. If the men decide to split the money according to the ratio of the amount of time each put into the project, how much will Artie get?

a. $3,000
b. $3,500
c. $4,000
d. $4,500

34. Tom's semi-weekly salary is $400. Jim's semi-monthly salary is $1,800. If both men work a standard 40-hour work week, which man earns more for the month of February? (Assume that this is NOT a leap year.)

a. Tom by $1,400
b. Jim by $400
c. Tom by $400
d. Jim by $1,400

35. Caleb can type 60 reports in 3 hours. Ethan can type 110 reports in 6 hours. Working together, how fast will it take them to type 375 reports?

a. 13 hours
b. 12 hours
c. 10 hours
d. 9 hours

36. For somebody who works a 30-hour work week, a $28,000 yearly salary translates into which of the following hourly wages?

a. $13.46
b. $14.50
c. $17.95
d. $19.46

Tank and Pipe Questions

37. A tank containing fluid is half full. A pipe that can fill $\frac{1}{16}$ of the tank per minute begins letting more fluid in. At the same time, a drain that can empty $\frac{1}{8}$ of the tank in one minute is opened. How long will it take to empty the tank?

a. 8 minutes
b. 16 minutes
c. 18 minutes
d. 32 minutes

38. Pipe T leads into a tank and Pipe V drains the tank. Pipe T can fill the entire tank in 6 minutes. Pipe V can drain the entire tank in 4 minutes. At a certain point in time, the valves leading to both pipes are shut and the tank is $\frac{1}{4}$ full. If both valves are opened simultaneously, how long will it take for the pipe to drain?

a. 2 minutes
b. 3 minutes
c. 4 minutes
d. 6 minutes

39. For every 10,000 liters of water that pass through a filtering system, 0.7 gram of a pollutant is removed. How many grams of the pollutant are removed when 10^6 liters have been filtered?

a. 7 grams
b. 70 grams
c. 700 grams
d. 7,000 grams

40. Rudy forgot to replace his gas cap the last time he filled his car with gas. The gas is evaporating out of his 14-gallon tank at a constant rate of $\frac{1}{3}$ gallon per day. How much gas does Rudy lose in 1 week?

a. 2 gallons
b. $2\frac{1}{3}$ gallons
c. $3\frac{1}{3}$ gallons
d. $4\frac{2}{3}$ gallons

41. Pipe A leads into a tank and Pipe B drains the tank. Pipe A can fill the entire tank in 10 minutes. Pipe B can drain the entire tank in 8 minutes. At a certain point in time, the valves leading to both pipes are shut and the tank is $\frac{1}{2}$ full. If both valves are opened simultaneously, how long will it take for the pipe to drain?

a. 18 minutes
b. 20 minutes
c. 22 minutes
d. 24 minutes

Distance Questions

42. A car travels at a constant rate of 60 km. per hour for 3 hours. How far did the car travel?

a. 180 kilometers
b. 180 miles
c. 18 kilometers
d. 18 miles

43. If Michael runs at a constant rate of 2.5 meters per second, how long will it take him to run 1 kilometer?

a. 4 minutes
b. 40 minutes
c. 400 seconds
d. 4,000 seconds

44. It took T.J. 20 minutes to jog 2 miles. What was his average speed in miles per hour?

a. 40 mph
b. 10 mph
c. 8 mph
d. 6 mph

45. Sipora drove to Stephanie's house at a constant rate of 45 mph. If Stephanie's house is 220 miles away and Sipora wants to get home in exactly 4 hours, how fast should she drive?

a. 50 mph
b. 55 mph
c. 60 mph
d. 65 mph

46. Amy can run 8 miles at a constant rate in 40 minutes. Sharon can run 12 miles at a constant rate in an hour. Who has a faster rate?

a. Amy
b. Sharon
c. They both run at the same rate.
d. It cannot be determined by the information given.

47. Train A travels at 60 mph for 20 minutes. Train B travels at 55 mph for 30 minutes. If both trains are traveling at a constant rate, which train would have traveled a greater distance after the time periods specified?

a. Train A
b. Train B
c. Both trains traveled the same distance.
d. It cannot be determined by the information given.

48. A train leaves a station traveling west at 60 mph. At the same time, another train heads east on a parallel track, traveling at a rate of 70 mph. If the 2 trains are initially 700 miles apart, how far apart are they after 1 hour?

a. 630 miles
b. 610 miles
c. 570 miles
d. 560 miles

49. Train A leaves Station A at 6 P.M., traveling east at a constant rate of 70 mph. At the same time, Train B leaves Station B, traveling west at a constant rate of 90 miles per hour. If the 2 trains pass each other at 8 P.M., then how far apart are the 2 stations?

a. 280 miles
b. 300 miles
c. 320 miles
d. 360 miles

50. An eastbound train destined for Stony Brook Station leaves Penn Station at 4 P.M., traveling at a rate of 60 miles per hour. At the same time, a westbound train departs the Stony Brook Station on its way to Penn Station. If the westbound train travels at a constant speed of 70 miles per hour and the two stations are 260 miles apart, at what time will the 2 trains pass each other?

a. 4:30 P.M.
b. 5:00 P.M.
c. 5:30 P.M.
d. 6:00 P.M.

ANSWERS

1. **a.** First, calculate the total amount of money: \$4,000 + \$3,500 + \$4,500 = \$12,000. He puts 30% of the \$12,000, or .30 × \$12,000 = \$3,600 into the checking account.
2. **c.** $\frac{1}{8}$ of the \$120 went to Suzanne: $\frac{1}{8} \times 120$ = \$15. This means there was 120 – 15 = \$105 left. $\frac{1}{4}$ of the \$105 went to Darlene: $\frac{1}{4} \times 105$ = \$26.25. Thus, the amount remaining is 105 – 26.25 = \$78.75.
3. **a.** $\frac{1}{3}$ of 12,000 = $\frac{1}{3} \times$ 12,000 = \$4,000 went to checking. $\frac{1}{4}$ of 12,000 = $\frac{1}{4} \times$ 12,000 = \$3,000 went to the CD. And $\frac{1}{8}$ of \$12,000 = $\frac{1}{8} \times$ 12,000 = \$1,500 went to buy the computer. Thus, the amount left equals 12,000 – 4,000 – 3,000 – 1,500 = \$3,500.
4. **b.** $2\frac{1}{2}$ feet = 2 feet 6 inches. $3\frac{1}{3}$ feet = 3 feet 4 inches. The sum of these values is 5 feet 10 inches.
5. **d.** Divide the total weight by 3 to figure out how much each of the 3 reels weigh: 6,720 ÷ 3 = 2,240 lbs each. Next, divide the weight of the reel by $\frac{3.2 \text{ lbs}}{\text{ft}}$: 2,240 lbs ÷ $\frac{3.2 \text{ lbs}}{\text{ft}}$ = 700 feet.
6. **c.** Five monitors will cost \$175 × 5 = \$875; Three printers will cost \$120 × 3 = \$360; Eight surge suppressors will cost \$18 × 8 = \$144. Before the discount, this adds up to: \$875 + \$360 + \$144 = \$1,379. 12% of \$1,379 = .12 × 1,379 = \$165.48. Thus, the final cost will be \$1,379 – 165.48 = \$1,213.52.
7. **c.** A 2:3 ratio is equivalent to an 8:12 ratio. Just multiply the $\frac{2}{3}$ ratio by $\frac{4}{4}$ to get $\frac{8}{12}$.
8. **b.** If 1 inch on the map denotes 1km, then 3 inches on the map would represent 3 kilometers.
9. **d.** First, set up a proportion: $\frac{27}{3} = \frac{x}{7}$. You can reduce the first fraction: $\frac{9}{1} = \frac{x}{7}$ and then cross-multiply: $1(x) = 9(7)$, so $x = 63$.
10. **b.** To find the distance Ralph can hike in 3 hours, first set up the ratio of the distance he can walk in a certain amount of time. 45 minutes is equal to $\frac{3}{4}$ of an hour or .75 hours. $\frac{1.3 \text{ miles}}{0.75 \text{ hours}}$. Then, set up the second ratio, $\frac{d}{3 \text{ hours}}$. Set these 2 ratios equal to each other. $\frac{1.3}{0.75} = \frac{d}{3}$.
11. **c.** First, set up a proportion: $\frac{18}{1} = \frac{x}{6}$. Cross-multiplying yields $18 \times 6 = 1 \times x$, and $x = 108$.
12. **d.** First, set up a proportion: $\frac{5}{3} = \frac{x}{18}$. Next, cross multiply: $3x = 18 \times 5$. Then, solve for your answer: $3x = 90$, so $x = 30$ minutes.
13. **b.** 9 to 5 represents an 8 hour work day, less the one hour lunch break yields 7 working hours. Multiply the 7 hours by 2.5 patients per hour = 17.5 patients. Of the choices, 18 patients is the best answer.
14. **a.** Set up a proportion: $\frac{1.5 \text{ cm}}{1 \text{ ft}} = \frac{12 \text{ cm}}{? \text{ ft}}$. Cross-multiply to get $1.5 \cdot ? = 12 \cdot 1$, or $1.5 \cdot ? = 12$. Divide both sides by 1.5 to get ? = 8 ft.
15. **b.** First, convert the height of the statue to inches. 305 ft. × 12 in. = 3,660 in. The statue is 3,660 + 1, or 3,661, inches tall. Next, set up a proportion: $\frac{1}{60} = \frac{x}{3,661}$. Cross multiply: $60x = 3,661$. Divide both sides by 60: $x = \frac{3,661}{60}$; x is about 61 inches. Convert to feet by dividing by 12: 61 ÷ 12 = 5 with a remainder of 1. Thus, the answer is 5 feet 1 inch, choice **b**.

16. **c.** Because this is a rate of work problem, consider what fraction of the job would get done in one minute. Scott would get $\frac{1}{30}$th of the job done while Henri would get $\frac{1}{60}$th of the job done in one minute. Together, they would get:

$$\frac{1}{30} + \frac{1}{60} = \frac{2}{60} + \frac{1}{60} = \frac{3}{60} = \frac{1}{20}$$

of the job done in one minute. Therefore, 20 minutes would be needed to pot 100 plants, and 40 minutes to pot all 200 plants.

17. **b.** Francine has read $\frac{4}{5}$ of 350 pages, or 0.8 × 350 = 280. Lydia has read half of that, or 140. Lydia has read 140 pages out of 350, or $\frac{140}{350}$. Reduce to $\frac{2}{5}$.

18. **d.** This is a two-step problem involving multiplication and simple subtraction. First, determine the amount of sand contained in the 4 trucks. $\frac{3}{4} \times \frac{4}{1} = \frac{12}{4}$. Next, reduce: $\frac{12}{4} = 3$. Finally, subtract: $3 - 2\frac{5}{6} = \frac{1}{6}$. There is $\frac{1}{6}$ ton more than is needed.

19. **c.** Semi-monthly means twice a month. This means he makes 2 × $1,200 = $2,400 per month. Multiply by 12 months per year: $\frac{12 \text{ months}}{\text{year}} \times \frac{\$2{,}400}{\text{months}} = \$28{,}800$ a year.

20. **c.** If he typically earns $800 a week, he makes $800 ÷ 40 hours = $20 per hour. This means he will make 1.5 × 20 = $30 for each overtime hour. 10 hours × $\frac{\$30}{\text{hour}} = \$300$.

21. **c.** First, add up all the hours he worked: $5 + 8 + 4\frac{1}{2} = 17\frac{1}{2}$ hours. Next, multiply the number of hours he worked by his hourly wage: 17.5 hrs × $\frac{\$22}{\text{hr}} = \$385$.

22. **a.** Subtract all of the listed deductions and diversion to yield the net paycheck: $2,300 – $160.45 – $82.50 – $73.25 – $100 = $1,883.80.

23. **b.** They can load 1 truck in the first 4 hours and $\frac{1}{2}$ a truck in the next 2 hours, so they can load $1\frac{1}{2}$ trucks in 6 hours.

24. **d.** First, if one machine assembles 400 parts in a half hour, it will assemble 800 parts in an hour. Two such machines working together will assemble 2 × 800 = 1,600 parts per hour. In 5 hours, they will assemble 5 × 1,600 = 8,000 parts. Of these 8,000 parts, 5% will be defective, so 95% will be non-defective. 95% of 8,000 = 95% × 8,000 = .95 × 8,000 = 7,600.

25. **d.** A daily salary is *per day*. She makes $120 per day times 24 days: $120 day × 24 days = $2,880.

26. **d.** First, add up all of his merchandise sales: $825 + $980 + $600 = $2,405. Next, take 8% of the $2,405: .08 × $2,405 = $192.40. Add the $192.40 commission to his 3 months of pay: $192.40 + (3)($1,600) = $192.40 + $4,800 = $4,992.40.

27. **b.** If 4 machines can complete the job in 6 hours, it will take 1 machine 4 times as long or 24 hours. It would take 3 machines $\frac{1}{3}$ of 24 hrs = $\frac{1}{3} \times 24 = 8$ hours.

28. **d.** If it takes 16 workers 10 days to complete a job, it would take 1 worker 16 times that amount, or 160 days. It would take 8 workers 160 ÷ 8 = 20 days. Also, notice that if the amount of workers is halved, the amount of time will be doubled.

29. **a.** It would take 1 worker 6 × 18 = 108 days. It would take 9 workers 108 ÷ 9 = 12 days.

30. **c.** It would take 1 person 9 × 12 = 108 days to complete the job. It would take 36 people 3 days to complete the same job because 108 ÷ 3 = 36.

31. **c.** Anthony can complete $\frac{1}{8}$ of the task in 1 hour. You know this because he completes the entire task in 8 hours. Together, Anthony and Dave complete $\frac{1}{3}$ of the task in 1 hour. (Thus, they are done in 3 hours). Convert both fractions into *twenty-fourths*. $\frac{8}{24}$ per hour (both men) – $\frac{3}{24}$ per hour (just Anthony) = $\frac{5}{24}$ per hour (just Dave). Thus, Dave completes $\frac{5}{24}$ of the task per hour. It will take him $\frac{24}{5}$ hours to complete the entire task. $\frac{24}{5} = 4\frac{4}{5}$ hours.

32. **d.** Here 40 hours of work + 60 hours of work = 100 total hours. Therefore, when considering the percent of work each did, it would be fair to give Rose 40% of the money and Marie 60% of the money. Marie gets 60% of $2,000, or 60% × $2,000 = .60 × $2,000 = $1,200. Alternatively, when combining their efforts, Marie and Rose earned a total of $2,000 for $100 of work. This is a rate of $20 per hour. Since Marie worked 60 hours, she gets 60 hrs × $\frac{\$20}{\text{hr}}$ = $1,200.

33. **c.** The ratio of time spent is 18:24 which reduces to 3:4. Use this 3 to 4 ratio in the algebraic equation $3x + 4x = 7x$, where $3x$ is the amount of money Al gets, $4x$ is the amount of money Artie gets, and $7x$ is the total amount of money (which we know is $7,000). Thus, if $7x$ = $7,000, x = $1,000. Artie's share equals $4x$ or (4)($1,000) = $4,000. Alternatively, you can calculate the fractional part of the job that each man worked and then use that fraction to calculate each man's share of the contracted amount. Al worked 18 hours and Artie worked 24 hours. The combined work time is 18 + 24 = 42 hours. This means the fractional part of the job for Al and Artie equals $\frac{18}{42}$ and $\frac{24}{42}$, respectively. Thus, Artie gets $\frac{24}{42}$ of the total $7,000. $\frac{24}{42}$ reduces to $\frac{4}{7}$. $\frac{4}{7}$ of $7,000 = $4,000, choice **c**.

34. **b.** Tom gets paid $400 semi-weekly (2 times a week) so he gets $800 per week. Multiply this weekly amount by the 4 weeks per month: $\frac{\$800}{\text{wk}} \times \frac{4\text{ wk}}{\text{mo}}$ = $3,200 per month. Jim gets paid $1,800 twice a month (semi-monthly) so he gets $3,600 per month. This means Jim makes $400 more per month than Tom does.

35. **d.** Ethan can type 110 reports in 6 hours, so he must type 55 reports in 3 hours. If Caleb types 60 reports and Ethan types 55 reports in 3 hours, the total number equals 125 reports. Now, compare this value with the 375 reports in the question. If they type 125 reports together in 3 hours, it will take them 3 times as long to type 375 reports. 3 hours × 3 = 9 hours, choice **d**.

36. **c.** The person works a 30-hour work week for $\frac{52\text{ weeks}}{\text{year}}$; $\frac{30\text{ hrs}}{\text{wk}} \times \frac{52\text{ wk}}{\text{yr}}$ = 1,560 hours. Next, divide the total amount of money by the total amount of hours: $28,000 ÷ 1,560 = $17.95 per hour.

37. **a.** Use *sixteenths* when considering the situation. This means $\frac{1}{16}$ is coming in as $\frac{1}{8} = \frac{2}{16}$ is going out. So, every minute the net loss of fluid is $\frac{2}{16} - \frac{1}{16} = \frac{1}{16}$ per minute loss. Since the tank starts out $\frac{1}{2}$ full, it is $\frac{8}{16}$ full. If $\frac{1}{16}$ drains per minute, it will take 8 minutes for the $\frac{8}{16}$ to drain.

38. **b.** Pipe T fills $\frac{1}{6}$ of the tank every minute. Pipe V empties $\frac{1}{4}$ of the tank per minute. This means the net effect every minute is $\frac{1}{4} - \frac{1}{6} = \frac{3}{12} - \frac{2}{12} = \frac{1}{12}$ of the tank is drained. If $\frac{1}{4}$ of the tank is initially full, this equals $\frac{3}{12}$ full. It will take 3 minutes for these $\frac{3}{12}$ to drain out at a rate of $\frac{1}{12}$ per minute.

39. **b.** 10,000 liters = 10^4 liters. Since 10^6 liters = 100 times 10^4, the number of grams of pollutant that is removed is 100 times 0.7, or 70.

40. **b.** $\frac{1}{3}$ gallon is lost per day over the course of a week, or 7 days. So, you multiply: $\frac{1}{3}$ gallons per day × 7 days = $\frac{7}{3}$ gallons, or $2\frac{1}{3}$ gallons are lost. Notice that it doesn't matter that the tank holds 14 gallons because the amount lost doesn't come close to 14.

41. **b.** Pipe A fills $\frac{1}{10}$ of the tank every minute. Pipe B empties $\frac{1}{8}$ of the tank per minute. This means the net effect every minute is $\frac{1}{8} - \frac{1}{10} = \frac{5}{40} - \frac{4}{40} = \frac{1}{40}$ of the tank is drained. If $\frac{1}{2}$ of the tank is initially full, this equals $\frac{20}{40}$ full. It will take 20 minutes for the $\frac{20}{40}$ to drain out at a rate of $\frac{1}{40}$ per minute.

42. **a.** Use the constant rate equation: $D = RT$. Here $D = \frac{60 \text{ km}}{\text{hr}} \times 3 \text{ hr} = 180 \text{ km}$.

43. **c.** 1 kilometer = 1,000 meters. Use $D = RT$ with $D = 1{,}000$, $R = \frac{2.5 \text{ m}}{\text{sec}}$, and T is the unknown. Rearrange $D = RT$ to $T = \frac{D}{R} = \frac{1{,}000}{2.5} = 400$ seconds.

44. **d.** Rearrange $D = RT$ into $R = \frac{D}{T}$. Substitute in the given values: $R = 20 \text{ min} = \frac{1}{3}$ hour, $D = 2$ mi into $R = \frac{D}{T}$ and $R = 2 \text{ mi} \div \frac{1}{3} \text{ hr} = 6$ mph.

45. **b.** Rearrange $D = RT$ to $R = D \div T = 220 \div 4 = 55$ mph.

46. **c.** Rearrange $D = RT$ into $R = D \div T$ by dividing both sides of the equation by T. Amy's rate is then $R = 8 \text{ mi} \div 40 \text{ min} = \frac{.2 \text{ mi}}{\text{min}}$. Next, calculate Sharon's rate in the same units of miles per minute. This means you need to convert the 1 hour into 60 min. Sharon's rate is then $R = 12 \text{ mi} \div 60 \text{ min} = \frac{.2 \text{ mi}}{\text{min}}$.

47. **b.** First, convert minutes to hours: 20 minutes = $\frac{1}{3}$ hour and 30 minutes = $\frac{1}{2}$ hour. Next, calculate the 2 distances by using $D = RT$. Train A will travel $D = 60 \times \frac{1}{3} = 20$ miles. Train B will travel $D = 55 \times \frac{1}{2} = 27.5$ miles. Thus, Train B travels the greater distance.

48. **c.** The first train will travel $D = RT = 60 \times 1 = 60$ miles west. The second train will travel $D = RT = 70 \times 1 = 70$ miles east. Thus, if the initial distance between the 2 trains was 700 miles, now the distance is 700 miles – 60 miles – 70 miles = 700 – 130 = 570 miles.

49. **c.** The total distance covered is equal to the distance that both trains travel. Train A travels east, a total of $D = RT = 70 \times 2 = 140$ miles. Train B travels west, a total of $D = RT = 90 \times 2 = 180$ miles. Note that $T = 2$ because the trains pass each other after 2 hours. Thus, the total initial distance is 140 miles + 180 miles = 320 miles.

50. **d.** The total distance will be equal to the distances traveled by both trains throughout the unknown amount of time (T).

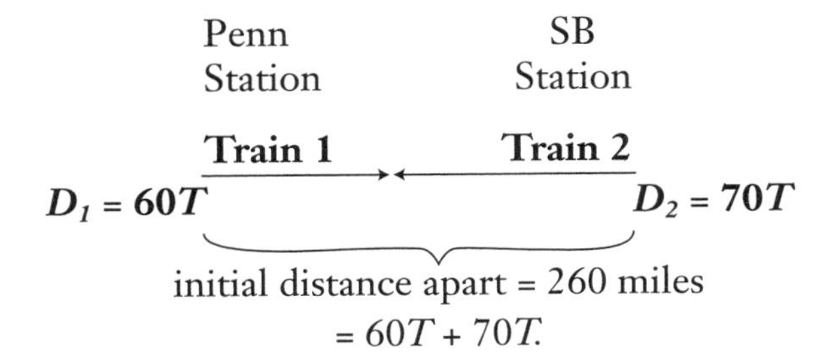

Thus, $260 = 60T + 70T = 130T$, and $T = 2$. The trains will pass each other after 2 hours, so the time will be 6:00 P.M., choice **d.**

Charts, Tables, and Graphs

When you pick up the newspaper or watch a news report on TV, you will often see information presented in a graph. More and more, we give and receive information visually. That's one reason you are likely to find graphs on math tests, and a good reason to understand how to read them. This chapter reviews the common kinds of graphs, charts, and tables you should be able to interpret.

PIE CHARTS

Pie charts show how the parts of a whole relate to one another. A pie chart is a circle divided into slices or wedges. Each slice represents a category. Pie charts are sometimes called circle graphs. Let's look at an example of a pie chart on the following page and see what kind of information it provides.

Example:

The pie chart below represents data collected from a recent telephone survey.

How Federal Dollars Are Spent

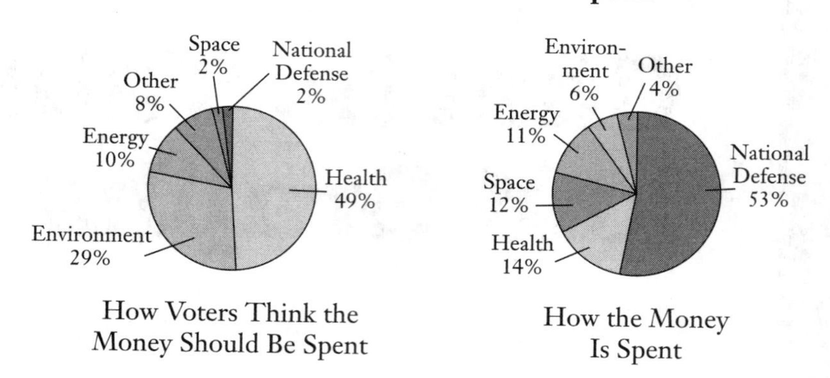

Using the "How Federal Dollars Are Spent" pie chart, answer the following questions:

1. Based on the survey, which category of spending best matches the voter's wishes?
2. On which category of spending did the voters want most of the money spent?
3. Which category of spending receives the most federal dollars?
4. To which two categories of spending did voters want the most money to go?
5. Which two categories of spending actually received the most money?

Explanations:

1. Energy: Voters say they would like about 10% of the budget to be spent on energy, and about 11% is actually spent on energy.
2. Health.
3. National defense.
4. Voters wanted money to go to health and environment.
5. Defense and health received the most money.

▶ LINE GRAPHS

Line graphs show how two categories of data or information (sometimes called *variables*) relate to one another. The data is displayed on a grid and is presented on a scale using a horizontal and a vertical axis for the different categories of information compared on the graph. Usually, each data point is connected together to form a line so that you can see trends in the data and how the data changes over time. Often you will see line graphs with *time* on the horizontal axis. Let's look at an example of a line graph and see the kind of information it can provide.

Example:
Consider the following information:

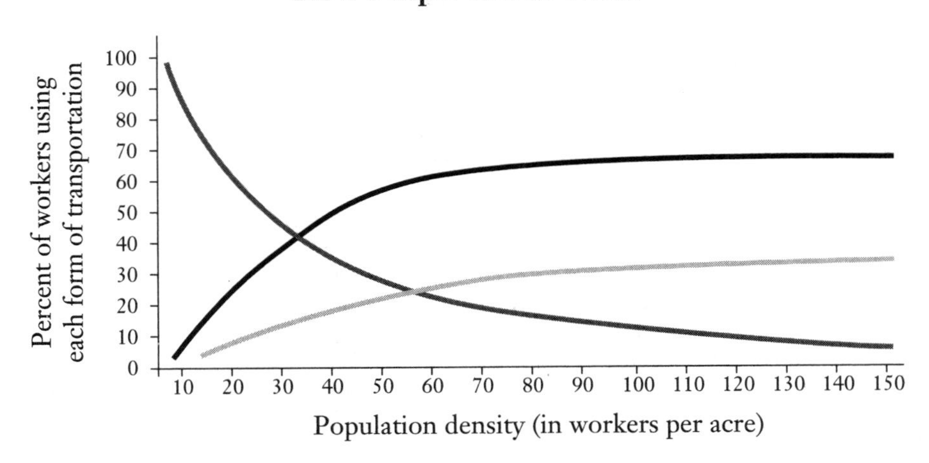

Using the "How People Get to Work" line graph, answer the following questions.

1. What variable is shown on the vertical axis? What variable is shown on the horizontal axis?
2. As the population density increases, will more or fewer people drive their own cars to work?
3. At about what point in population density does the use of public transportation begin to level off?
4. Which form of transportation becomes less popular as population density increases?

Explanations:

1. Look at the labels. The percent of workers using each form of transportation is shown on the vertical axis. Population density is shown on the horizontal axis.
2. As population density increases, less people use their own cars to get to work.
3. At about 80 to 100 workers per acre, the percentage of workers using public transportation begins to level off at about 70%.
4. Find the line that moves down as population density increases. It's the line labeled "own car." This is the form of transportation that decreases as population density increases.

BAR GRAPHS

Like pie charts, *bar graphs* show how different categories of data relate to one another. A bar represents each category. The length of the bar represents the relative frequency of the category—compared to the other categories on the graph. Let's look at an example of a bar graph on the next page and see the kind of information it can provide.

Example:

The following bar graph compares the 2002 rainfall amounts in Cherokee County with the average rainfall in Cherokee County over the last five years.

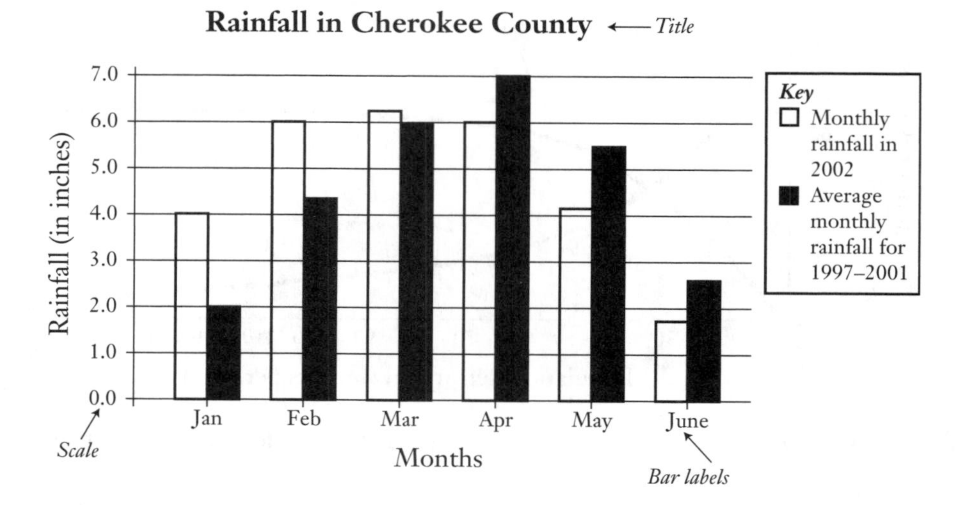

Using the "Rainfall in Cherokee County" bar graph, answer the following questions.

1. What does each bar represent? What is the difference between the shaded bars and the white bars?
2. During which months is the rainfall in 2002 greater than the average rainfall?
3. During which months is the rainfall in 2002 less than the average rainfall?
4. How many more inches of rain fell in April 2002 than in January 2002?
5. How many more inches of rain fell in January 2002 than on average during the last five years in January?

Explanations:

1. Look at the labels and the key. Each bar represents the number of inches of rainfall during a particular month. From the key, you know that the shaded bars represent the average monthly rainfall for 1996–2001. The white bars represent the rainfall in 2002.
2. Compare the white bars with the shaded bars. Rainfall in 2002 is greater than average during the months that the white bar is taller than the shaded bar for that month. Rainfall in 2002 was greater than the average rainfall during January, February, and March.
3. Compare the white bars with the shaded bars. Rainfall in 2002 is less than average during the months that the shaded bar is taller than the white bar for that month. Rainfall in 2002 was less than the average rainfall during April, May, and June.
4. Compare the height of the white bars for January and April. In April, 6 inches of rain fell. In January, 4 inches of rain fell. Then subtract: $6 - 4 = 2$. So, in April, 2 more inches of rain fell than in January.

5. Compare the height of the shaded bar and the white bar for January. The shaded bar represents 2 inches. The white bar represents 4 inches. Subtract: 4 – 2 = 2. So, two more inches of rain fell in January 2002 than on average during the last five years in January.

GETTING INFORMATION FROM TABLES

Tables present information in rows and columns. Rows go across, or horizontally. Columns go up and down, or vertically. The box, or cell, that is made where a row and a column meet provides specific information. When looking for information in tables, it's important to read the table title, the column headings, and the row labels so you understand all of the information. Let's look at some examples of tables and the types of information you might expect to learn from them.

THE FUJITA-PEARSON TORNADO INTENSITY SCALE

CLASSIFICATION	WIND SPEED (IN MILES PER HOUR)	DAMAGE
F0	72	Mild
F1	73–112	Moderate
F2	113–157	Significant
F3	158–206	Severe
F4	207–260	Devastating
F5	261–319	Cataclysmic
F6	320–379	Overwhelming

Example:

Using the "Fujita-Pearson Tornado Intensity Scale" table, answer the following questions.

1. If a tornado has a wind speed of 173 miles per hour, how would it be classified?
2. What kind of damage would you expect from a tornado having a wind speed of 300 miles per hour?
3. What wind speed would you anticipate if a tornado of F6 were reported?

Explanations:

1. F3. The wind speed for F3 tornados ranges from 158–206 mph.
2. Cataclysmic: F5 tornados range in wind speed of 261–319 mph and cause *cataclysmic* damage.
3. F6 tornados range from wind speeds of 320–379 miles per hour.

PRACTICE QUESTIONS

Use the chart below to answer questions 1 through 5.

NAME	SCORE
Darin	95
Miguel	90
Anthony	82
Christopher	90
Samuel	88

1. What is the mean score of the people listed?
 a. 90
 b. 89
 c. 88
 d. 85

2. What is the median score of the people listed?
 a. 90
 b. 89
 c. 88
 d. 85

3. What is the range of the scores listed?
 a. 90
 b. 50
 c. 24
 d. 13

4. What is the mode of the scores listed?
 a. 90
 b. 89
 c. 88
 d. 85

5. If Anthony's score was incorrectly reported as an 82 when his actual score on the test was a 90, which of the following statements would be true when his actual score is used in the calculations?

a. the mean, median, range, and mode will change.
b. the mean, median, and range, will change; the mode will remain the same.
c. only the mean and median will change.
d. none of the above.

6. The chart below gives the times that 4 swimmers had in their race. Which swimmer had the fastest time?

SWIMMER	TIME (SEC)
Molly	38.51
Jeff	39.23
Asta	37.95
Risa	37.89

a. Molly
b. Jeff
c. Asta
d. Risa

The chart below lists the number of members present at the monthly meetings for the Environmental Protection Club. Use this chart to answer questions 7 through 9.

MONTH	# OF MEMBERS
September	54
October	61
November	70
December	75

7. What was the average monthly attendance over the course of all the months listed?

a. 71
b. 65
c. 61
d. 56

8. What was the median value of members in attendance during the course of the four months shown?

a. 54
b. 61
c. 65.5
d. 70

9. If the data presented in the table were plotted as a bar graph, which of the following would best represent the data most accurately?

a.

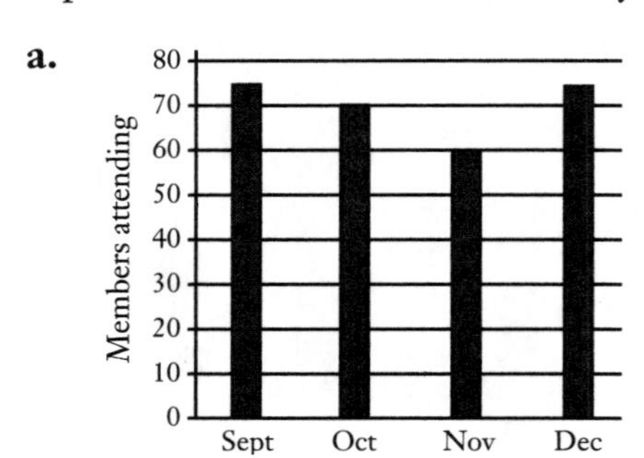

b.

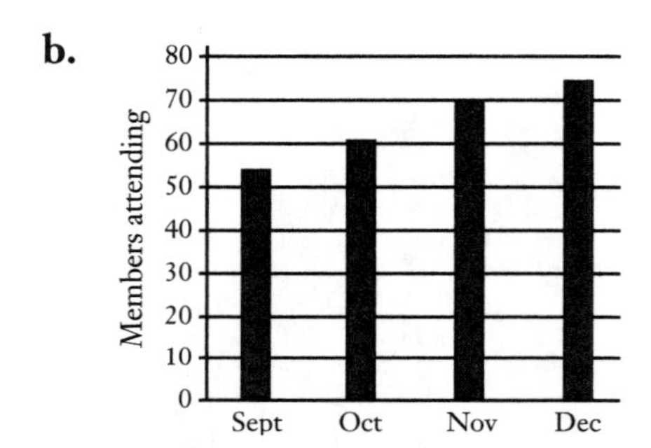

c.

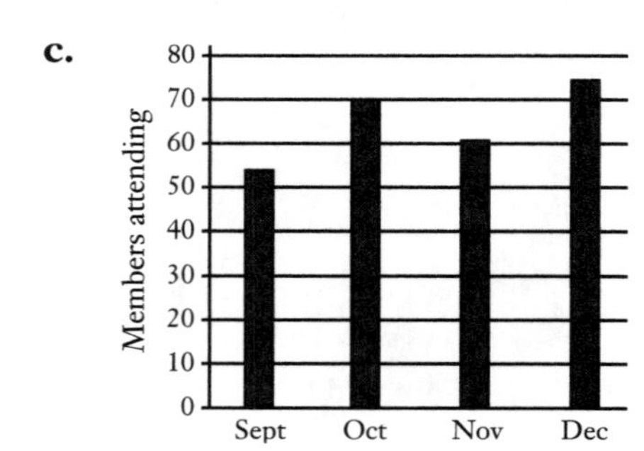

d.

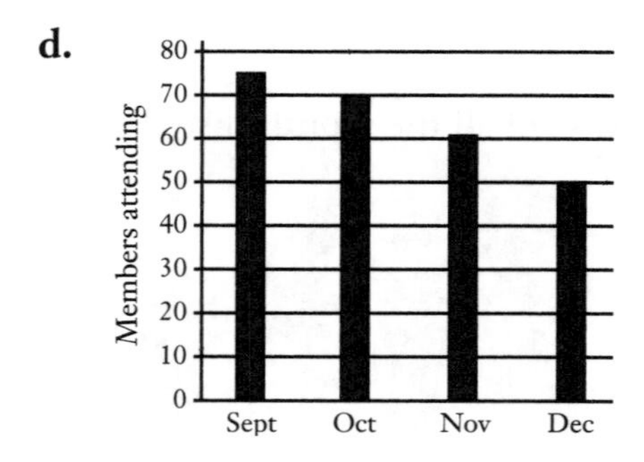

The pie chart below shows the Johnson family budget for one month. Use this information to answer questions 10 through 12.

Johnson Family Budget

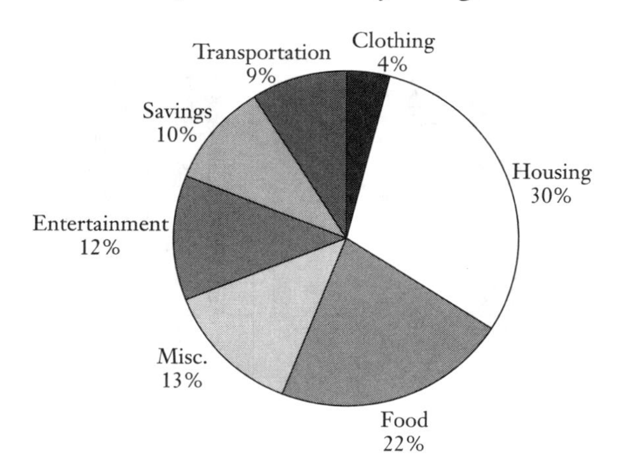

10. In percent of overall expenses, how much more money is spent on food than on transportation and clothing combined?

a. 9%
b. 11%
c. 13%
d. 22%

11. If the Johnson family budget is $4,000 per month, how much money is spent on housing each month?

a. $800
b. $1,000
c. $1,200
d. $1,400

12. If the Johnson family budget is $4,000 per month, how much money will they save each year?

a. $48,000
b. $4,800
c. $400
d. none of the above

The graph below shows the yearly electricity usage for Finnigan Engineering Inc. over the course of three years for three departments. Use this information to answer questions 13 through 16.

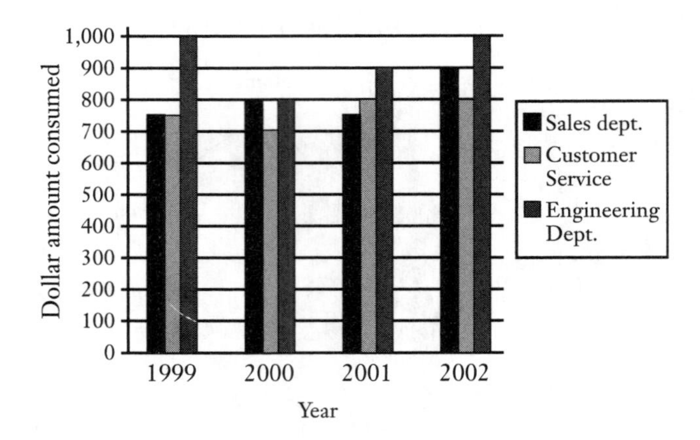

13. How much greater was the electricity cost for Sales during the year 1999 than the electricity cost for Customer Service in 2000?

a. $200
b. $150
c. $100
d. $50

14. Which of the following statements is supported by the data?

a. The Sales Department showed a steady increase in the dollar amount of electricity used during the 4-year period.
b. The Customer Service Department showed a steady increase in the dollar amount of electricity used during the 4-year period.
c. The Engineering Department showed a steady increase in the dollar amount of electricity used from 2000–2002.
d. none of the above

15. What was the percent decrease in electricity usage (in dollar amount) from 1999 to 2000 for the Engineering Department?

a. 25%
b. 20%
c. 15%
d. 10%

16. If the information in the bar graph associated with question 13 is transcribed and a line graph is generated, which of the following line graphs is correct?

a.

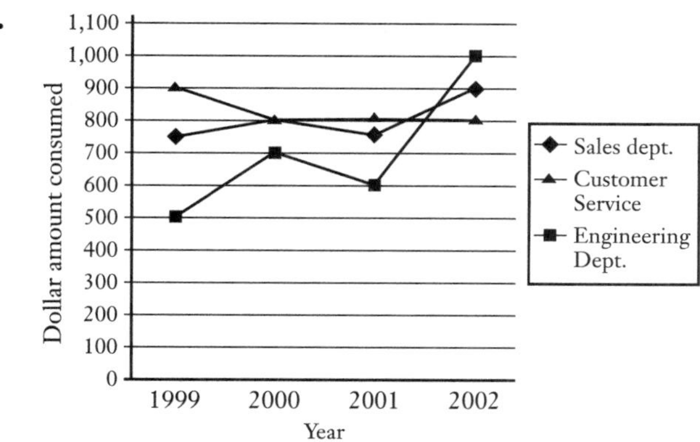

b.

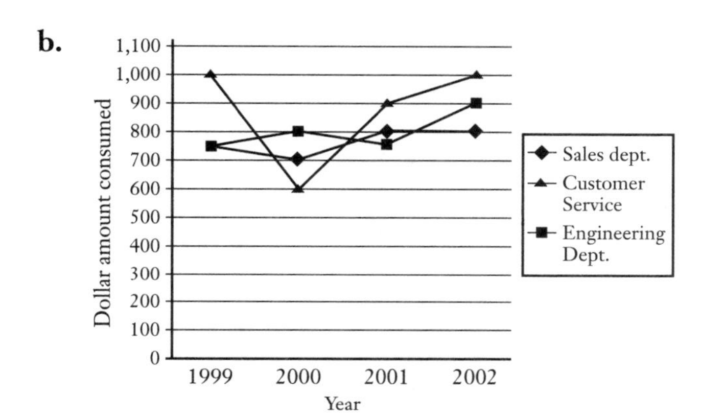

c.

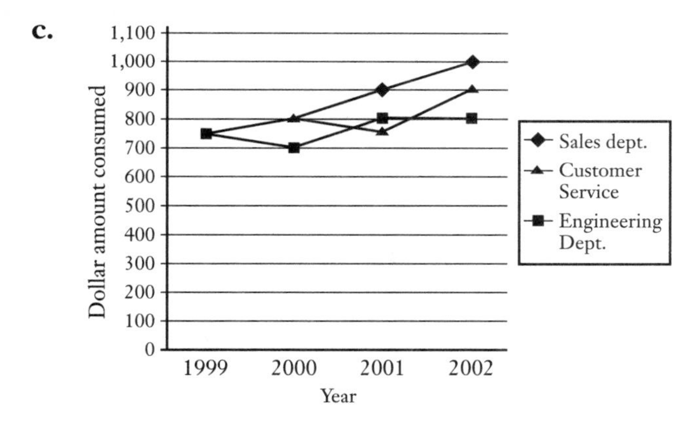

d.

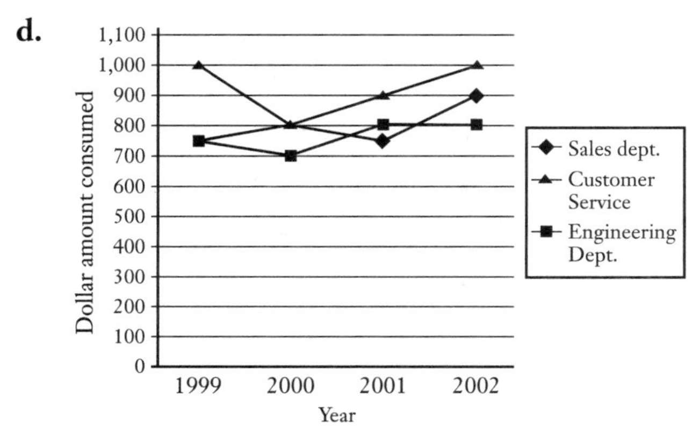

The table below shows the numbers of male and female students involved in several school activities. Use this information to answer questions 17–19.

ACTIVITY	MALE	FEMALE
Drama	11	13
Journalism	12	10
Science Club	9	11
Debate	12	15

17. Which activity has the lowest ratio of males to females?
  a. Drama
  b. Journalism
  c. Science Club
  d. Debate

18. For all of the students listed, what percent of the students are involved in Debate?
  a. 15%
  b. 20%
  c. 27%
  d. 29%

19. If 3 more males and 4 more females join the Science Club, what percent of the students will be in this club?
  a. 15%
  b. 20%
  c. 27%
  d. 29%

Use the chart below to answer questions 20 through 23.

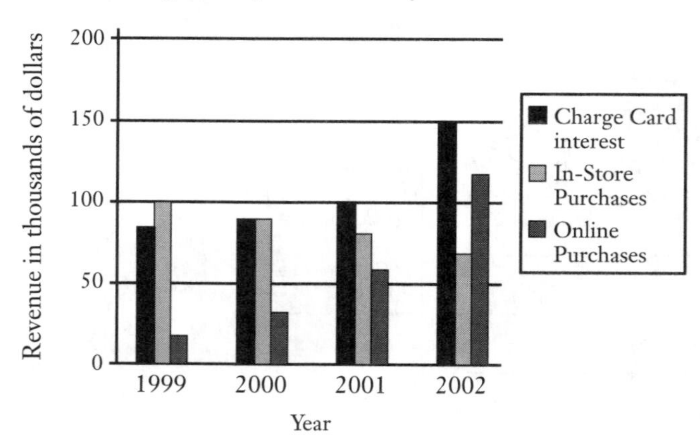

20. Based on the chart above, which answer choice represents a true statement?

a. Online Purchases have increased whereas Charge Card Interest has decreased over the course of the 4 years shown.

b. Charge Card Interest has increased whereas Online Purchases have decreased over the course of the 4 years shown.

c. In-Store Purchases have increased whereas Charge Card Interest has decreased over the course of the 4 years shown.

d. Online Purchases have increased whereas In-Store Purchases have decreased over the course of the 4 years shown.

21. If all of the information on the graph above were converted into a table, which of the following tables would correctly display the data (with revenue in thousands of dollars)?

a.

	1999	2000	2001	2002
Charge Card Interest	$90	$90	$100	$150
In-Store Purchases	$80	$90	$80	$70
Online Purchases	$15	$60	$30	$120

b.

	1999	2000	2001	2002
Charge Card Interest	$80	$90	$100	$120
In-Store Purchases	$80	$80	$80	$70
Online Purchases	$15	$60	$60	$120

c.

	1999	2000	2001	2002
Charge Card Interest	$80	$90	$100	$150
In-Store Purchases	$100	$90	$80	$70
Online Purchases	$15	$30	$60	$120

d.

	1999	2000	2001	2002
Charge Card Interest	$80	$90	$100	$150
In-Store Purchases	$90	$80	$90	$70
Online Purchases	$15	$30	$60	$120

22. The Online Purchases in 1999 were what fraction of the Charge Card Interest in 2002?

a. $\frac{1}{5}$

b. $\frac{1}{10}$

c. $\frac{1}{4}$

d. $\frac{1}{2}$

23. In-Store Purchases in 1999 made how much more than In-Store Purchases in 2002?

a. $30

b. $60

c. $6,000

d. $30,000

The line graph below shows earning for the three divisions of Steinberg Lumber Company throughout the 4 quarters in 2002. Use the information presented to answer questions 24 through 26.

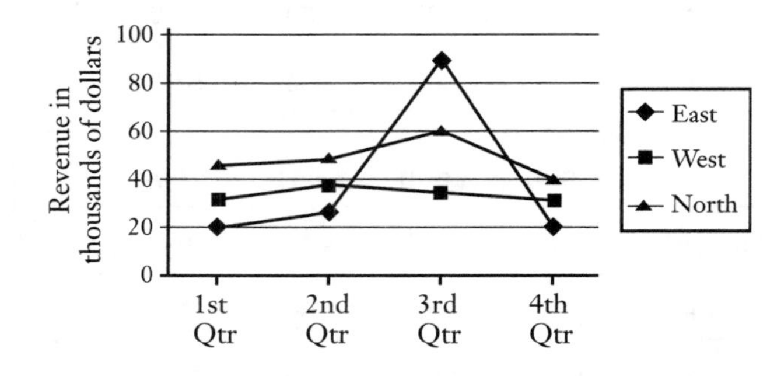

24. Which of the following statements is true?

a. The East Division consistently brought in more revenue than the other 2 divisions.

b. The North Division consistently brought in more revenue than the West Division.

c. The West Division consistently out performed the East Division.

d. Both **b** and **c** are true.

25. What is the percent decrease in revenue for the North Division when analyzing dollar amounts from the 3rd and 4th quarters?

a. $33\frac{1}{3}\%$

b. 40%

c. 50%

d. 60%

26. During the year 2002, Steinberg Lumber secured a major contract with a developer in Canada. The East and North Divisions both supplied lumber for this project. Which of the following statements seems to be supported by the data?

a. The West Division was angry that the other two divisions supplied the lumber for this contract.
b. The next big contract will be covered by the West Division.
c. The contract with the Canadian developer was secured in the third quarter.
d. The contract with the Canadian developer was secured in the fourth quarter.

Use the information below to answer questions 27 through 29. The pie chart shows the percentage of employees in the various departments of Amelia Computer Consultants Inc.

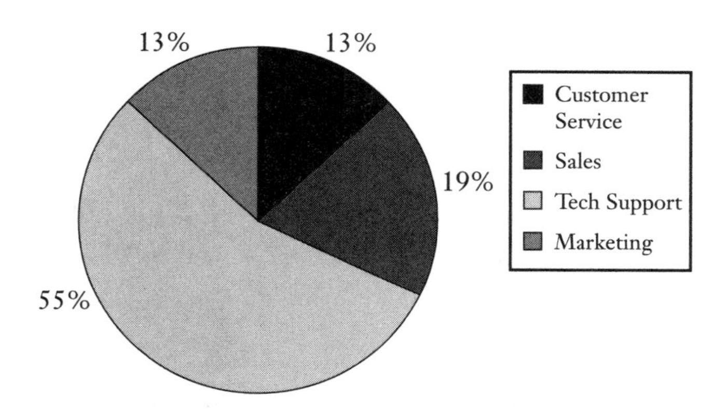

27. Which two departments account for 32% of the employees?

a. Marketing and Tech Support
b. Customer Service and Sales
c. Sales and Tech Support
d. Marketing and Customer Service

28. If the total number of employees is 400, how many employees are in the Tech Support department?

a. 52
b. 76
c. 110
d. 220

29. Suppose that the Customer Service department is expanded by adding 12 new employees. Which of the following statements is true?

a. Customer Service and Marketing have the same number of employees.
b. The percent of employees in Marketing is now 11%.
c. The percent of employees in sales is now 20%.
d. The percent of employees in Tech Support is now 53%, while the percent of employees in Customer Service is 16%.

Use the information below to answer questions 30 and 31. The chart below shows the composition by percent of the human body with respect to various elements.

30. If a man weighs 260 pounds, how much does the carbon in his body weigh?

ELEMENT	PERCENT BY WEIGHT
Carbon	18%
Hydrogen	10%
Oxygen	65%
Other Elements	7%

a. 46.8 pounds
b. 48.6 pounds
c. 52.4 pounds
d. 54.2 pounds

31. The chart below shows the cost for different categories of UTP cabling. If Athena's office needs to buy 100 feet of UTP cable that can send data at a speed of 75 megabytes per second, about how much will she spend?

CATEGORY	CHARACTERISTICS	PRICE PER FOOT
Category 1	Does not support data transmission	$.75
Category 2	Supports data transmission speeds up to 4 megabytes per second	$ 1.00
Category 3	Supports data transmission speeds up to 16 megabytes per second	$ 1.75
Category 4	Supports data transmission speeds up to 20 megabytes per second	$ 2.50
Category 5	Supports data transmission speeds up to 100 megabytes per second	$ 3.00

a. $3
b. $250
c. $275
d. $300

32. During the year 2000, at Deluxe Vacuum Co., the East and West divisions had equal sales and the North sold the most. Which graph could be the graph of Deluxe's yearly sales for 2000?

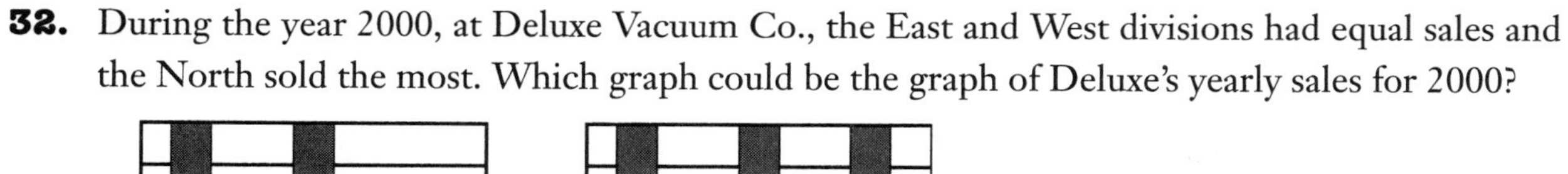

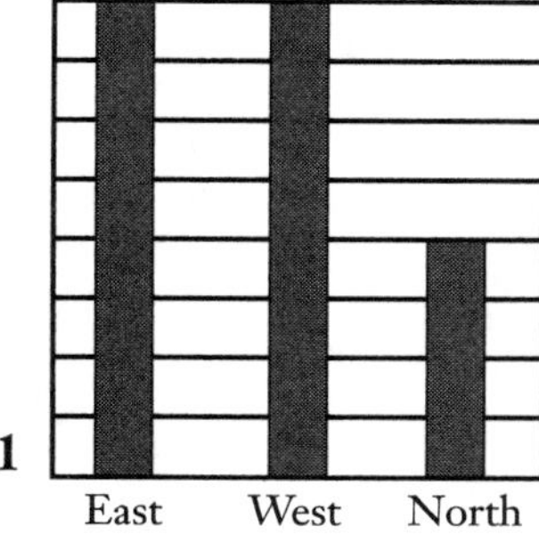

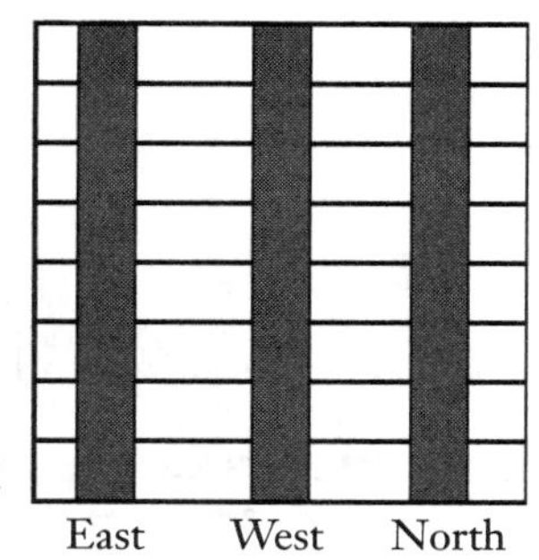

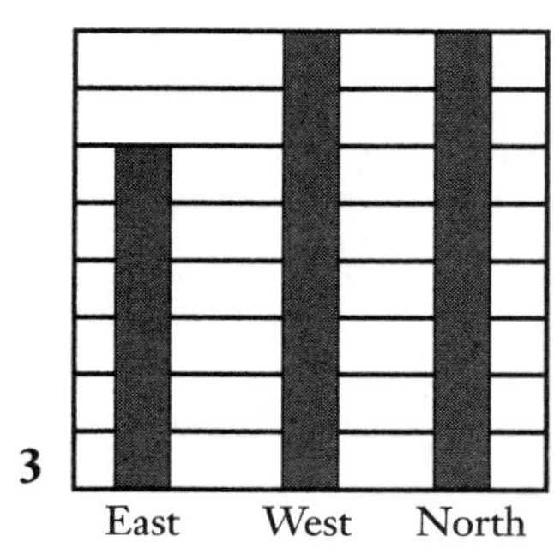

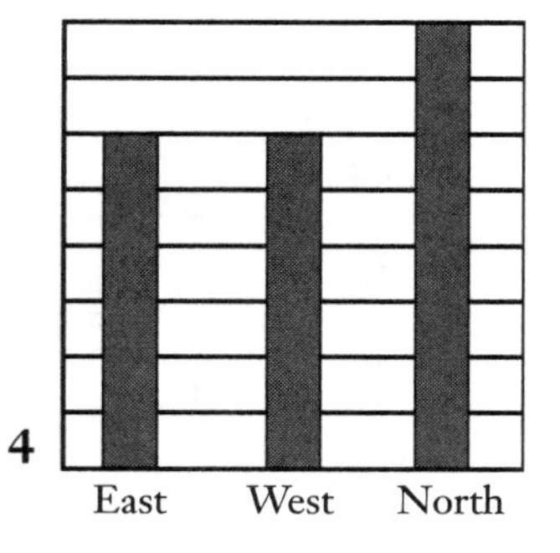

a. 1
b. 2
c. 3
d. 4

Use the following information to answer questions 33 and 34. Swimming Pool World pledged to donate 3.2% of their sales during the second week of May to the Children's Hospital. Below is their sales chart for May.

MAY SALES

Week 1	$5,895
Week 2	$73,021
Week 3	$54,702
Week 4	$67,891

33. How much did Swimming Pool World donate to the children's hospital?
a. $2,336.67
b. $3,651.05
c. $23,366.72
d. $36,510.50

34. If Swimming Pool World pledged 1% of sales for the entire month of May, how much would they have donated?

- **a.** about $300 more
- **b.** about $300 less
- **c.** about $500 more
- **d.** about $500 less

35. The chart below shows registration for art classes for Fall 2003.

STUDENTS REGISTERING FOR ART CLASSES

Course	Number of Students
Stained Glass	21
Beginning Drawing	48
Sculpture	13
Watercolors	18
TOTAL	100

If this is a representative sampling, how many out of 500 students would be expected to choose stained glass for their art course?

- **a.** 21
- **b.** 92
- **c.** 105
- **d.** 210

Use the following information to answer questions 36 through 37. The following table shows the rainfall, in inches, over a 5-day period in August for Hilo, Hawaii. It also includes the total rainfall for the year and the average rainfall for a typical year.

	RAINFALL	YEAR	NORMAL
Monday	0.08	90.88	79.15
Tuesday	0.09	90.97	79.16
Wednesday	0.70	91.67	79.17
Thursday	0.19	91.86	79.17
Friday	0.32	92.18	79.50

36. Find the average rainfall for the 5-day period in August.

a. 1.38 inches
b. 0.276 inches
c. 0.32 inches
d. 0.237 inches

37. Using Monday's reading and rounding off to the nearest one percent, the year-to-date record is what percent of the normal reading?

a. 13%
b. 15%
c. 87%
d. 115%

Use the following information to answer questions 38 through 40. The chart below shows the colors of replacement parts for pocket PCs. The total number of parts shipped is 1,650.

BOXED SET OF REPLACEMENT PARTS

Part Color	Number of Pieces
Green	430
Red	425
Blue	
Yellow	345
TOTAL	1,650

38. If a person randomly grabbed a part out of the box, what is the probability that the part will be blue?

a. $\frac{1}{4}$
b. $\frac{1}{9}$
c. $\frac{1}{12}$
d. $\frac{3}{11}$

39. Approximately what percent of the total shipment is red?

a. 18%
b. 20%
c. 26%
d. 30%

40. If the chart below shows the number of replacement parts that were found to be defective, what percent of the new parts is defective?

BOXED SET OF REPLACEMENT PARTS

Part Color	Number of Defective Pieces
Green	14
Red	10
Blue	8
Yellow	12

a. $22\frac{1}{3}\%$
b. 18%
c. $8\frac{1}{2}\%$
d. $2\frac{2}{3}\%$

Use the following information to answer questions 41 through 43. The table lists the size of building lots in the Orange Grove subdivision and the people who are planning to build on those lots. For each lot, installation of utilities costs $12,516. The city charges impact fees of $3,879 per lot. There are also development fees of 16.15 cents per square foot of land.

LOT	AREA (SQ. FT.)	BUILDER
A	8,023	Ira Taylor
B	6,699	Alexis Funes
C	9,004	Ira Taylor
D	8,900	Mark Smith
E	8,301	Alexis Funes
F	8,269	Ira Taylor
G	6,774	Ira Taylor

41. The area of the smallest lot listed is approximately what percent of the area of the largest lot listed?

a. 25%
b. 50%
c. 75%
d. 85%

42. How much land does Mr. Taylor own in the Orange Grove subdivision?
  a. 23,066 sq. ft.
  b. 29,765 sq. ft.
  c. 31,950 sq. ft.
  d. 32,070 sq. ft.

43. How much will Mr. Smith pay in development fees for his lot?
  a. $1,157.00
  b. $1,437.35
  c. $143,735
  d. $274,550

44. Felipe is planning to get an Internet service in order to have access to the World Wide Web. Two service providers, A and B, offer different rates as shown in the table below. If Felipe plans on using 25 hours of Internet service per month, which of the following statements is true?

INTERNET SERVICE RATES

Provider	Free Hours	Base Charge	Hourly Charge
A	17.5	$20.00	$1.00
B	20	$20.00	$15.0

  a. Provider A will be cheaper.
  b. Provider B will be cheaper.
  c. The providers will cost the same per month.
  d. The answer cannot be determined from the information given.

45. Refer to the table below to answer this question: If you take recyclables to the recycler who will pay the most, what is the greatest amount of money you could get for 2,200 pounds of aluminum, 1,400 pounds of cardboard, 3,100 pounds of glass, and 900 pounds of plastic?

RECYCLER	ALUMINUM	CARDBOARD	GLASS	PLASTIC
X	$.06/pound	$.03/pound	$.07/pound	$.02/pound
Y	$.07/pound	$.04/pound	$.08/pound	$.03/pound

  a. $409
  b. $440
  c. $447
  d. $485

46. Which of the following brands is the least expensive?

BRAND	PRICE ($)	WEIGHT (OZ.)
W	0.21	6
X	0.48	15
Y	0.56	20
Z	0.96	32

a. W
b. X
c. Y
d. Z

Use the following information to answer questions 47 through 50.

When an earthquake occurs, some of the energy released travels through the ground as waves. Two general types of waves are generated. One type is called the P wave, and the other is called the S wave. A graph can be made of the travel times of these waves.

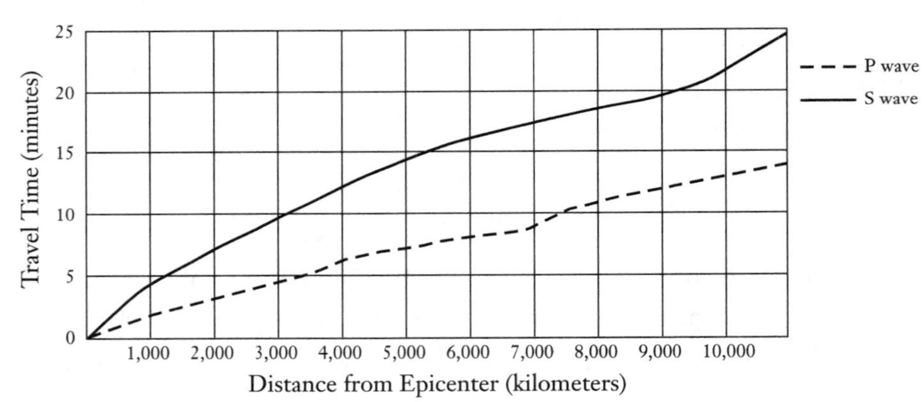

47. How many minutes does it take the S wave to travel 5,500 kilometers?
a. 15 min.
b. 20 min.
c. 25 min.
d. 30 min.

48. Approximately how many minutes does it take a P wave to travel 8,000 km?
a. 6 min.
b. 12 min.
c. 3 min.
d. 15 min.

49. An earthquake occurs at noon, and the recording station receives the S wave at 12:04 P.M. How far away is the earthquake?

a. 1,000 km.
b. 2,000 km.
c. 3,000 km.
d. 4,000 km.

50. How far away is an earthquake if the difference in arrival time between the P and S waves is 5 minutes?

a. 1,000 km.
b. 3,000 km.
c. 4,000 km.
d. 7,000 km.

ANSWERS

1. **b.** The formula for calculating the mean (average) is:

   $Mean = \frac{\textit{sum of all values}}{\textit{\# of values}}$.

   The sum of all the values given is: 95 + 90 + 82 + 90 + 88 = 445. The number of values (scores) is 5. Thus, the mean = $\frac{445}{5}$ = 89.
2. **a.** First, list all of the scores in order: 82, 88, 90, 90, 95. The middle score will be the *median*, thus 90 is the median.
3. **d.** The range is calculated by subtracting the lowest score from the highest score. Thus, the range is 95 – 82 = 13.
4. **a.** The *mode* is the score that occurs the most. Here, there are two nineties, thus 90 is the mode.
5. **d.** Calculate the new median, mode, and range and compare them to the original values. To find the new mean, first add all the scores: 95 + 90 + 90 + 90 + 88 = 453, and then divide by 5: 453 ÷ 5 = 90.6. Next, we can calculate the median and see if it is different: 88, 90, 90, 90, 95. Here, we see that the median is the same as it was before, 90. The mode is still 90 because 90 is the score that occurs the most. The range is now 95 – 88 = 7. The chart below compares the old and new values:

	OLD	NEW
Mean	89	90.6
Median	90	90
Mode	90	90
Range	13	95 – 88 = 7

   Thus, choice **d** is the correct answer.
6. **d.** The fastest swimmer will have the quickest time. 37.89 (Thirty seven and eighty-nine hundredths is the fastest). Thus, Risa is the fastest swimmer.
7. **b.** The formula for calculating the mean (average) is:

   $Mean = \frac{\textit{sum of all values}}{\textit{\# of values}}$.

   The sum of all the values given is: 54 + 61 + 70 + 75 = 260. The number of values is 4. Thus, the mean = 260 ÷ 4 = 65.
8. **c.** List all of the values in order: 54, 61, 70, 75. Here, there is an even number of values, so we average the middle 2 numbers. The average of 61 and 70 is $\frac{131}{2}$ = 65.5.
9. **b.** The number of members attending for the four months was: 54, 61, 70, 75, for September, October, November, and December, respectively. This is accurately displayed in choice **b.** Note that choice **b** is also the only choice that depicts the ascending trend. That is to say, the number of members in attendance increases over time.

10. **a.** 22% is spent on food. When we combine transportation (9%) and clothing (4%), the sum is 13%. Thus, the amount spent on food is 22% – 13% = 9% greater.

11. **c.** Housing consumes 30% of the monthly budget. 30% of $4,000 is calculated by multiplying: 30% × $4,000 = .30 × $4,000 = $1,200.

12. **b.** They save 10% of $4,000 each month: .10 × $4,000 = $400. Over the course of a year they will save $400 per month × 12 months = $4,800.

13. **d.** The Sales Dept. (black bar) spent $750 on electricity in 1999. The Customer Service Dept. (lightest-colored bar) spent $700 on electricity in 2000. Thus, the Sales Dept. spent $750 – $700 = $50 more.

14. **c.** The usage for the Engineering Department increases by $100 each year from 2000 through 2002. None of the other statements are supported by the data. Claims of steady increase over the course of 4 years would be visually represented as 4 bars, each with greater height than the prior.

15. **b.** The difference in dollar amounts used is $1,000 – $800 = $200. When compared with the original $1,000 consumed, this can be expressed as a percent by equating $\frac{200}{1,000} = \frac{x}{100}$. Thus, x = 20%.

16. **d.** The line graph in choice **d** accurately displays the data that is obtained from the bar graph. The data is listed below in table format so that you can easily see the information present on both the bar graph and the correct line graph:

	1999	**2000**	**2001**	**2002**
Sales	750	800	750	900
Customer Service	750	700	800	800
Engineering	1,000	800	900	1,000

17. **d.** The M:F (male to female) ratios are as follows:

Drama: $\frac{11}{13} \approx .85$

Journalism: $\frac{12}{10} = 1.2$

Science Club: $\frac{9}{11} \approx .82$

Debate: $\frac{12}{15} = .8$

Here, .8 is the least value, so a $\frac{12}{15}$ ratio is the smallest M:F ratio listed.

18. **d.** This question is easily solved by adding a column and row labeled "TOTAL" onto the side and bottom of the given chart:

ACTIVITY	MALE	FEMALE	TOTAL
Drama	11	13	24
Journalism	12	10	22
Science Club	9	11	20
Debate	12	15	27
TOTAL			93

Now, you can easily see that 27 students out of the 93 total are taking debate. $\frac{27}{93} \approx .29$. To write these values as a percent, simply move the decimal point two places to the right and add the percent symbol: 29%.

19. **c.** Using the new information, our chart becomes:

ACTIVITY	MALE	FEMALE	TOTAL
Drama	11	13	24
Journalism	12	10	22
Science Club	12	15	27
Debate	12	15	27
TOTAL			100

This means that 27 out of 100 students are now in the Science Club. $\frac{27}{100} = 27\%$.

20. **d.** The black bars (Charge Card Interest) increase from year to year. The white bars (In-Store Purchases) decrease from year to year. The gray bars (Online Purchases) increase from year to year. Thus, only choice **d** is correct.

21. **c.** The black bars (Charge Card Interest) increase from 80 to 90 to 100 to 150. The white bars (In-Store Purchases) decrease from 100 to 90 to 80 to 70. The gray bars (Online Purchases) increase from 15 to 30 to 60 to 120. Only choice **c** presents this data correctly.

22. **b.** In 1999, Online Purchases were at \$15,000. In 2002, Charge Card Interest totaled \$150,000. Since 15 is $\frac{1}{10}$ of 150, the answer is $\frac{1}{10}$, choice **b.**

23. **d.** Note that all dollar amounts in the chart are expressed as, "Revenue in thousands of dollars." In 1999, the In-Store Purchases were at \$100,000. In 2002, the amount is \$70,000. Thus, the difference is \$30,000. Thus, choice **d**, \$30,000, is correct.

24. **b.** Looking at the graph, we see that the line for *North* (the line with triangular points) is always higher than the line for *West* (the line with the square points). All other statements are NOT supported by the data in the graph. Thus, only choice **b** is true.

25. **a.** Here, the revenue in thousand of dollars decreases from 60 to 40. Thus, the difference is 20. As compared with the original 60, this represents $\frac{20}{60}$ = .333 . . . . To express this as a percent, just move the decimal point 2 places to the right .3333 ➔ $33\frac{1}{3}$%.

26. **c.** Since we are told that this was a "major" contract, the statement best supported by the data is choice **c**: "The contract with the Canadian developer was secured in the third quarter." The data supports this statement because both the East and North Divisions had a significant revenue increase during the third quarter, which might be indicative of having a large contract for that quarter.

27. **b.** Customer Service (black) accounts are 13% of the total, and Sales (dark gray) accounts are 19% of the total. Together, these add to 32%. Since both Marketing and Customer Service are at 13%, either department could be combined with Sales to total 32% of the company employees. Note that only Customer Service and Sales are listed as a choice.

28. **d.** Tech Support (white) is 55% of the total. 55% of 400 equals 55% × 400 = .55 × 400 = 220. You can save time when answering a question like this by noticing that 55% will be slightly more than $\frac{1}{2}$ the total of 400, so slightly more that 200. Only choice **d** makes sense.

29. **d.** Before the addition of the 12 new customer service representatives, the number of employees in each department was as follows:

Customer Service: .13 × 400 = 52
Marketing: .13 × 400 = 52
Sales: .19 × 400 = 76
Tech Support: .55 × 400 = 220

The new total is 400 + 12 = 412. The new amount of customer service employees is 52 + 12 = 64. The percentages are as follows:

Customer Service: $\frac{64}{412} \approx .15534 \approx 15.5\% \approx 16\%$

Marketing: $\frac{52}{412} \approx .12621 \approx 12.6\% \approx 13\%$

Sales: $\frac{76}{412} \approx .18447 \approx 18.4\% \approx 18\%$

Tech Support: $\frac{220}{412} \approx .53398 \approx 53.4\% \approx 53\%$

Thus, the only choice that is true is choice **d**.

30. **a.** Carbon accounts for 18% of body weight. 18% of 260 = .18 × 260 = 46.8 pounds.

31. **d.** Since she needs to support a speed of 75 megabytes per second, only Category 5 UTP cable can be used. Note that Category 5 "*Supports data transmission speeds up to 100 megabytes per second.*" This cable cost $3 per foot, so 100 feet will cost 100 × $3.00 = $300.

32. **d.** The East and West divisions had equal sales, so we need a graph where the bars for East and West are the same height. North sold the most, so we need a graph that also shows North as having the largest bar in the graph. Graph 4 shows this situation. Thus, choice **d** is correct.

33. **a.** During Week 2 they made $73,021. To find 3.2% of this amount, just multiply by .032: .032 × $73,021 = $2,336.672. Rounded to the nearest cent, the answer is: $2,336.67.

34. **b.** First, calculate the total by adding up all the dollar amounts:

$$\begin{array}{r} \$5{,}895 \\ \$73{,}021 \\ \$54{,}702 \\ +\ \$67{,}891 \\ \hline \$201{,}509 \end{array}$$

Next, take 1% of the total by multiplying by .01.
$.01 \times \$201{,}509 = \$2{,}015.09$. This is about \$300 less than the \$2,336.67 that they actually donated.

35. **c.** Since the sampling is representative, this means that the same trend will be seen when a larger sample is considered. Thus, simply multiply by 5 to see how many students out of 500 will choose stained glass. $5 \times 21 = 105$.

36. **b.** Add up the values for the 5 days shown: $.08 + .09 + .70 + .19 + .32 = 1.38$. Divide this amount by 5 to get the average: $1.38 \div 5 = .276$ inches.

37. **d.** On Monday, the year-to-date record is 90.88 inches. The normal amount is 79.15. Thus the year-to-date value is obviously above 100% of the normal value, making choice **d** the only possible correct answer. (Note that $\frac{90.18}{79.15} \approx 1.1482 \approx 114.82\% \approx 115\%$.)

38. **d.** $430 + 425 + 345 = 1{,}200$ parts are accounted for. Since the total is 1,650, $1{,}650 - 1{,}200 = 450$ blue parts. When randomly picking a part, the chance of getting blue is 450 out of 1,650 $= \frac{450}{1{,}650}$. Simplify the expression: $\frac{450}{1{,}650} \div \frac{150}{150} = \frac{3}{11}$.

39. **c.** 425 out of 1,650 is red. $\frac{450}{1{,}650} = 425 \div 1{,}650 = .25\overline{757}$. To convert to a percent, just move the decimal point 2 places to the right and add the percent symbol: $25.7575\ldots\% \approx 26\%$.

40. **d.** Add a row for the total at the bottom of the given chart:

BOXED SET OF REPLACEMENT PARTS

Part Color	Number of Defective Pieces
Green	14
Red	10
Blue	8
Yellow	12
TOTAL DEFECTIVE	44

44 parts out of 1,650 are defective. $\frac{44}{1{,}650} = .02\overline{666}$. To express this as a percent, move the decimal point 2 places to the right and add the percent symbol: $2.66666\ldots\%$. This equals $2\frac{2}{3}\%$.

41. **c.** The smallest lot is 6,699 ft^2, and the largest lot is 9,004 ft^2. 6,699 out of 9,004 equals $\frac{6{,}699}{9{,}004} \approx .74400 \approx 74.40\% \approx 74\%$. Thus, choice **c**, 75% is the best approximation.

42. d. Look at the chart to see all of the land Mr. Taylor owns:

LOT	AREA (SQ. FT.)	BUILDER
A	8,023	Ira Taylor
B	6,699	Alexis Funes
C	9,004	Ira Taylor
D	8,900	Mark Smith
E	8,301	Alexis Funes
F	8,269	Ira Taylor
G	6,774	Ira Taylor

The total amount of land Mr. Taylor owns is 8,023 + 9,004 + 8,269 + 6,774 = 32,070 ft^2.

43. b. Mr. Smith's lot is 8,900 ft^2. You are told, "... There are also development fees of 16.15 cents per square foot of land." 16.15 cents = \$0.1615. Thus, he must pay \$.1615 × 8,900 = \$1,437.35 in development fees.

44. c. When used for $\frac{25 \text{ hrs}}{\text{mo}}$, Provider A will cost: \$20 plus 7.5 × \$1 (for the hourly charge above the free hours). This equals \$27.50. Provider B will cost \$20 plus 5 × \$1.50 (for the hourly charge above the free hours). This equals \$20 + \$7.50 = \$27.50 as well, so choice **c** is the correct answer.

45. d. Since Recycler Y pays more per pound for all 4 types of recyclables, all 4 items should be brought there. The aluminum will yield .07 × 2,200 = \$154. The cardboard will yield .04 × 1,400 = \$56. The glass will yield .08 × 3,100 = \$248. The plastic will yield .03 × 900 = \$27. These add to \$485.

46. c. Calculate the price per ounce (oz.) for each brand:

W: $\frac{.21}{6} = .035$

X: $\frac{.48}{15} = .032$

Y: $\frac{.56}{20} = .028$

Z: $\frac{.96}{32} = .03$

Thus, brand Y is the least expensive, choice **c**.

47. **a.** The solid line represents the S wave. This crosses 5,500 km at time = 15 minutes.

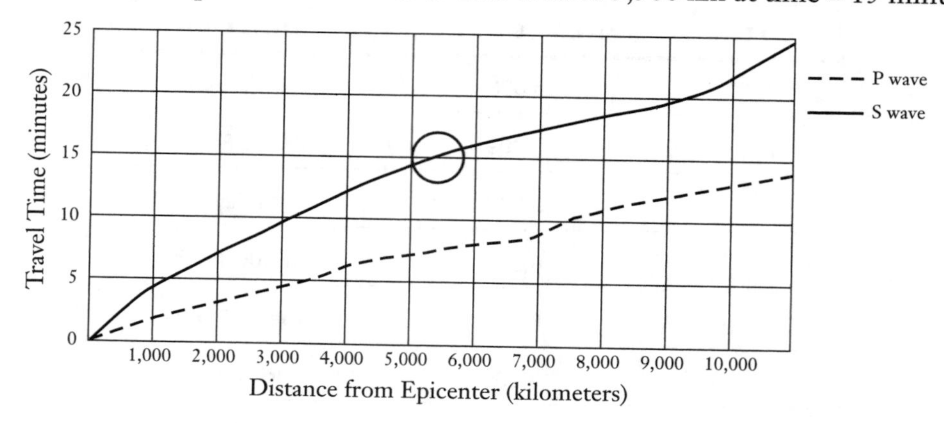

48. **b.** The P wave is the dashed line. It travels 8,000 km at a point above time = 10, but below time = 15. Hence, a time of 12 minutes is the best answer.

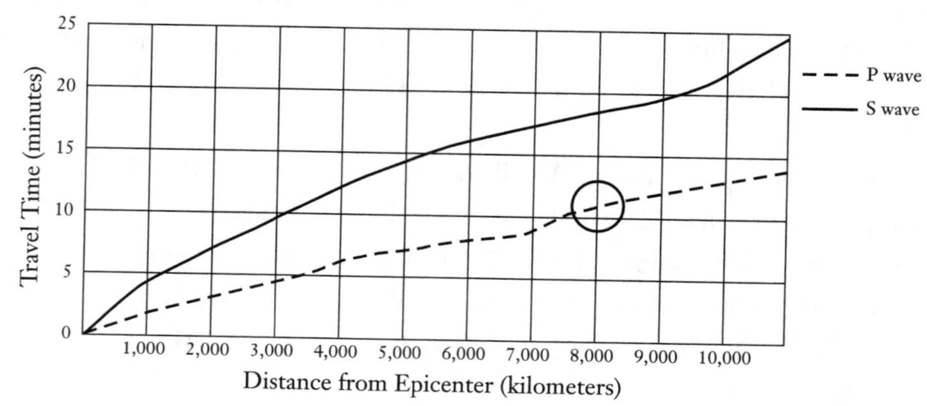

49. **a.** The S wave was received 4 minutes after the earthquake. Locate 4 minutes on the vertical axis of the graph and then move across until you reach the S wave graph. Look down to the horizontal axis to see that this means the earthquake is 1,000 km away.

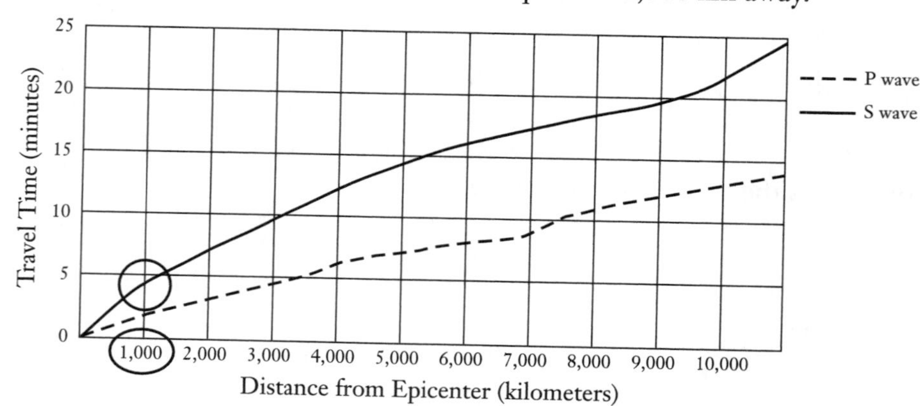

50. **c.** A difference in time of 5 minutes can be seen by looking at the vertical axis. The vertical axis is marked by 5-minute intervals, so use this distance to judge where the distance (gap) between the waves is also 5 minutes (see gap circled below). Look down to see the horizontal axis to note that this time difference occurs at 4,000 km.

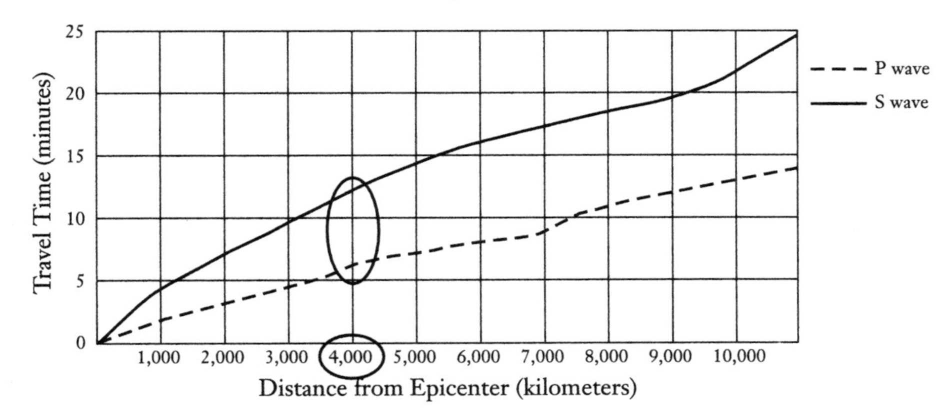

Geometry and Measurement

UNITS OF MEASUREMENT

Metric System

In the metric system lengths are calculated in meters, masses are calculated in grams, and volumes are calculated in liters. The prefix of each unit is very important. You should be familiar with the following prefixes:

PREFIX	MEANING	EXAMPLE
milli	$\frac{1}{1,000}$ of	1 milligram is $\frac{1}{1,000}$ of a gram
centi	$\frac{1}{100}$ of	1 centimeter is $\frac{1}{100}$ of a meter
deci	$\frac{1}{10}$ of	1 decigram is $\frac{1}{10}$ of a gram
deca	10 times	1 decameter is 10 meters
hecto	100 times	1 hectoliter is 100 liters
kilo	1,000 times	1 kilometer is 1,000 meters

Customary Units

The relationships between the customary units are not as systematic as the relationships between units in the metric system. Here, lengths are measured in inches, feet, yards, and miles. Weights are measured in pounds and ounces. And volumes are measured in cubic inches, cubic feet, and so forth. Below is a chart of common conversions for customary units.

COMMON CONVERSIONS

1 foot = 12 inches	1 cup = 8 fluid ounces
3 feet = 1 yard	1 pint = 2 cups
1 mile = 5,280 feet	1 quart = 2 pints
1 acre = 43,560 square feet	1 gallon = 4 quarts
1 ton = 2,000 pounds	1 pound = 16 ounces
1 gross = 144 units	1 liter = 1,000 cubic centimeters

▶ CONVERTING UNITS

Conversion factors are an easy way to convert units. For example, using the knowledge that 12 in. = 1 foot, you can generate 2 conversion factors: $\frac{12 \text{ in.}}{1 \text{ ft.}}$ and $\frac{1 \text{ ft.}}{12 \text{ in.}}$. Suppose you wanted to convert 5 feet into inches. You can use the conversion factor $\frac{12 \text{ in.}}{1 \text{ ft.}}$:

$$5 \not{\text{ft.}} \times \frac{12 \text{ in.}}{1 \not{\text{ft.}}} = 60 \text{ in.}$$

Notice that you crossed out the units you *didn't* want (feet) and ended up with the units you *did* want (inches). Having the feet in the denominator of this conversion factor lets us cross-out the "ft." unit in the original 1 ft. In other instances you may want to cross-out inches and convert to feet. The conversion factor to use would be $\frac{1 \text{ ft.}}{12 \text{ in.}}$.

Sample question:

32,000 ounces is equal to how many tons?

a. 16
b. 8
c. 4
d. 1

You know that 1 lb. = 16 oz. and 1 ton = 2,000 lbs. Use this information to make a series of conversion factors and multiply:

$32{,}000 \text{ oz.} \times \frac{1 \text{ lb.}}{16 \text{ oz.}} \times \frac{1 \text{ ton}}{2{,}000 \text{ lb.}} = 32{,}000 \cancel{\text{oz.}} \times \frac{1 \cancel{\text{lb.}}}{16 \cancel{\text{oz.}}} \times \frac{1 \text{ ton}}{2{,}000 \cancel{\text{lb.}}} = 1 \text{ ton}$. Thus, the correct answer is **d.** Notice that your goal is to cross-out the units you DO NOT want and to end up with the units that you DO want.

▶ CALCULATIONS WITH GEOMETRIC FIGURES

Perimeter is the distance around a figure. The perimeter of a circle is called its *circumference*. *Area* is a measure of the surface of a two-dimensional figure. *Volume* is a measure of the amount of space inside a three-dimensional shape. You should be familiar with the following formulas.

Triangle: Area = $\frac{1}{2}bh$

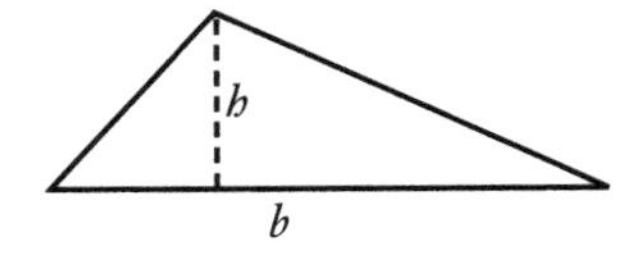

The interior angles of a triangle add to 180°.

The interior angles of a quadrilateral (4-sided polygon) add to 360°.

Square: Area = s^2

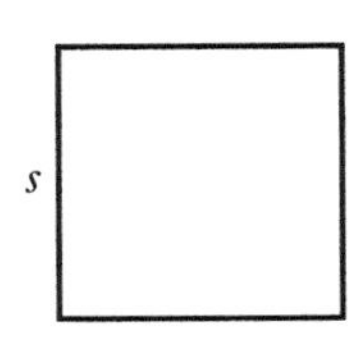

Rectangle: Area = lw

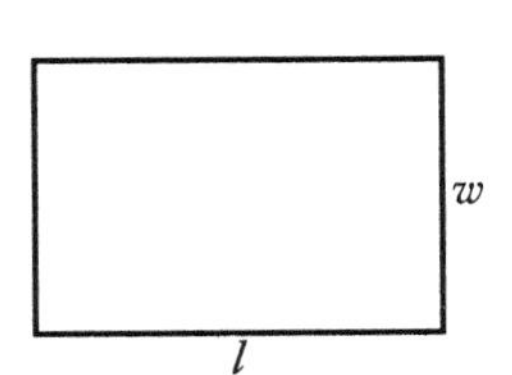

Circle: Area = πr^2

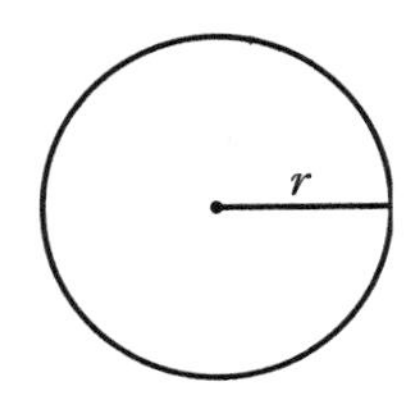

Circumference = $\pi d = 2\pi r$

($\pi \approx 3.14$ or $\frac{22}{7}$)

Parallelogram: Area = bh

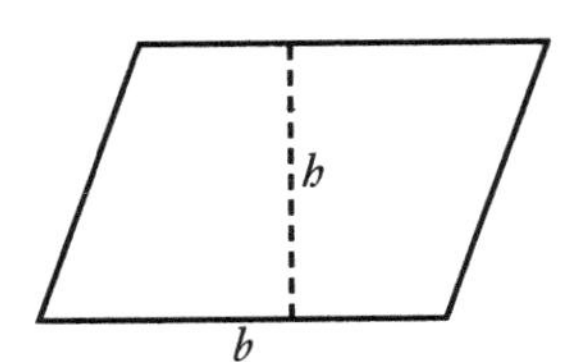

Trapezoid: Area = $\frac{1}{2}h(b_1 + b_2)$

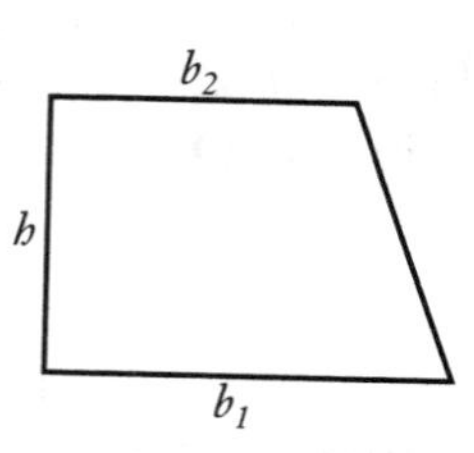

Pythagorean theorem: $a^2 + b^2 = c^2$

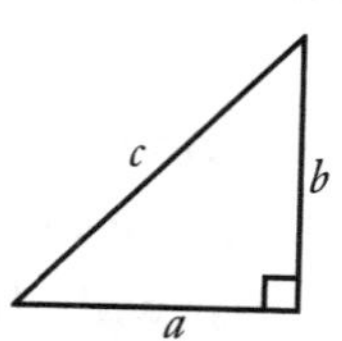

Right Circular Cylinder: Volume = $\pi r^2 h$

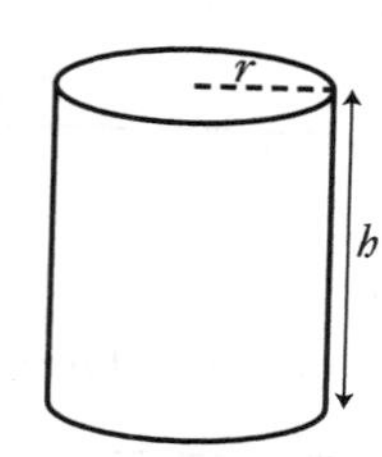

Total Surface Area = $2\pi rh + 2\pi r^2$

Rectangular Solid: Volume = lwh

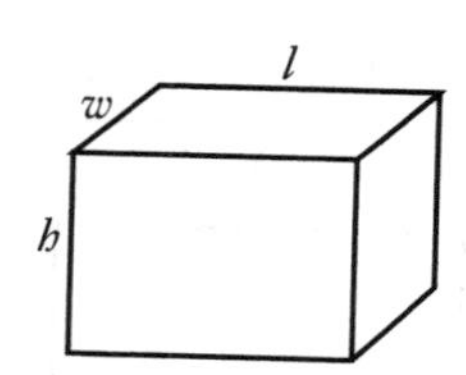

Total Surface Area = $2(lw) + 2(hw) + 2(lh)$

Sample Questions:

1. A rectangular swimming pool measures 204 feet long and 99 feet wide. What is the area of the pool in square yards?

a. 20,196 square yards
b. 6,732 square yards
c. 2,244 square yards
d. 1,800 square yards

The answer is **c.** Convert both the length and the width into yards:

$204 \text{ ft.} \times \frac{1 \text{ yd.}}{3 \text{ ft.}} = 68 \text{ yd.}$

$99 \text{ ft.} \times \frac{1 \text{ yd.}}{3 \text{ ft.}} = 33 \text{ yd.}$

Next, use the area formula for a rectangle, $A = lw$:

$A = 68 \text{ yd} \times 33 \text{ yd} = 2{,}244$ square yards.

2. One cubic centimeter of wood weighs 6 grams. How much would a cube weigh if it measured 10 cm on each side?

a. 60 grams
b. 600 grams
c. 6,000 grams
d. 60,000 grams

The answer is **c**. For this question, you are told that the weight is 6 grams per cubic centimeter, or $\frac{6\text{ g}}{\text{cm}^3}$. You need to find out how many cm^3 there are in the bigger cube, which is the *volume* of the cube. Recall that for a cube, $V = side^3$. The bigger cube has a side = 10, so $V = 10^3 = 1{,}000\text{ cm}^3$. Then, to find the weight you multiply $1{,}000\text{ cm}^3 \times \frac{6\text{ g}}{\text{cm}^3} = 6{,}000$ grams. Thus, choice **c** is correct.

PRACTICE QUESTIONS

1. What is the sum of 3 ft. 5 in., 10 ft. 2 in., and 2 ft. 7 in.?

a. 14 ft. 14 in.
b. 15 ft. 11 in.
c. 15 ft. 13 in.
d. 16 ft. 2 in.

2. Three pieces of pipe measure 5 ft. 8 in., 4 ft. 7 in., and 3 ft. 9 in. What is the combined length of all three pipes?

a. 14 ft.
b. 13 ft. 10 in.
c. 12 ft. 9 in.
d. 12 ft. 5 in.

3. How many inches are there in $3\frac{1}{3}$ yards?

a. 126 in.
b. 120 in.
c. 160 in.
d. 168 in.

4. 76,000 mL is equivalent to how many liters?

a. 7.6 L
b. 76 L
c. 760 L
d. 7,600 L

5. 2,808 inches is equivalent to how many yards?

a. 234
b. 110
c. 78
d. 36

6. What is the sum of 5 yd. 2 ft., 8 yd. 1 ft., 3 yd. $\frac{1}{2}$ ft., and 4 yd. 6 in.?

a. 20 yd. $\frac{1}{2}$ ft.
b. 20 yd. 1 ft.
c. 21 yd. 1 ft.
d. 21 yd. $\frac{1}{2}$ ft.

7. How many yards are in a mile?

a. 1,760
b. 4,400
c. 5,280
d. 63,360

Use the chart below to answer questions 8 through 10:

CUSTOMARY UNITS—METRIC UNIT CONVERSIONS

LENGTH
1 in. = 2.54 cm.
1 yard = .9 m.
1 mi. = 1.6 km.

8. Convert 3 ft. 5 in. into centimeters.

a. 104.14 cm.
b. 65.6 cm.
c. 51.3 cm.
d. 16.14 cm.

9. 5,500 yd. is equivalent to how many meters?

a. 13,970 m.
b. 11,400 m.
c. 9,800 m.
d. 4,950 m.

10. 1,280 miles is equal to how many kilometers?

a. 800 km.
b. 1,152 km.
c. 2,048 km.
d. 3,200 km.

11. A child has a temperature of 40 degrees *C*. What is the child's temperature in degrees Fahrenheit? ($F = \frac{9}{5}C + 32$)

a. 101°
b. 102°
c. 103°
d. 104°

12. If John was waiting for 45 minutes for an appointment with a contractor that lasted 1 hour and 25 minutes, what is the total amount of time spent at the contractor's office?

a. 2 hr. 10 min.
b. 2 hr. 25 min.
c. $2\frac{1}{2}$ hr.
d. 3 hr. 10 min.

13. There are 12 yards of twine on a roll. Danielle cuts off 2 feet of twine for a project. How many *feet* of twine are left on the roll?

a. 2 ft.
b. 34 ft.
c. 36 ft.
d. 142 ft.

Use the conversion chart below to answer questions 14 through 17:

LIQUID MEASURE
8 oz. = 1 c.
1 pt. = 2 c.
1 qt. = 2 pt.
4 qt. = 1 gal.

14. How many ounces are in 2 pints?

a. 16 oz.
b. 32 oz.
c. 44 oz.
d. 64 oz.

15. 364 oz. is equivalent to how many quarts?

a. 182 qt.
b. 91 qt.
c. 22.75 qt.
d. 11.375 qt.

16. How many ounces are in 3 gallons?

a. 384 oz.
b. 192 oz.
c. 96 oz.
d. 48 oz.

17. A 25-gallon tub of fluid will be poured into containers that hold half of a quart. If all of the containers are filled to capacity, how many will be filled?

a. 50
b. 100
c. 200
d. 250

18. A rotating door, pictured below, has 4 sections, labeled *a*, *b*, *c*, and *d*. If section *a* is making a 45 degree angle with wall 1, what angle is section *c* making with wall 2?

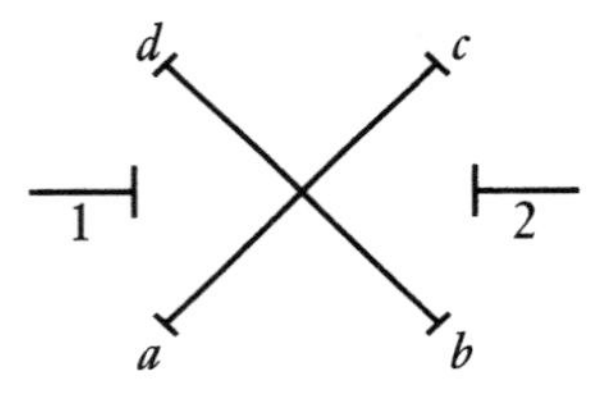

a. 15 degrees
b. 45 degrees
c. 55 degrees
d. 90 degrees

19. A rectangle has 2 sides equaling 6 ft and 1 yd, respectively. What is the area of the rectangle?

a. 6 ft^2
b. 12 ft^2
c. 18 ft^2
d. 20 ft^2

20. A square with $s = 6$ cm. has the same area of a rectangle with $l = 9$ cm. What is the width of the rectangle?

a. 4 cm.
b. 6 cm.
c. 8 cm.
d. 9 cm.

21. If the area of a circle is 9π cm^2, what is the circumference?

a. 3π cm.^{2}
b. 3π cm.
c. 6π cm.^{2}
d. 6π cm.

22. A rectangular tract of land measures 860 feet by 560 feet. Approximately how many acres is this? (1 acre = 43,560 square feet.)

a. 12.8 acres
b. 11.06 acres
c. 10.5 acres
d. 8.06 acres

23. Marguerite is redoing her bathroom floor. Each imported tile measures $1\frac{2}{7}$ in. by $1\frac{4}{5}$ in. What is the area of each tile?

a. $1\frac{8}{35}$ square inches
b. $1\frac{11}{35}$ square inches
c. $2\frac{11}{35}$ square inches
d. $3\frac{3}{35}$ square inches

24. A rectangular swimming pool measures 160 feet long and 80 feet wide. What is the perimeter of the pool in yards?

a. 40 yards
b. 160 yards
c. 240 yards
d. 280 yards

25. In the diagram, the angle x equals how many degrees?

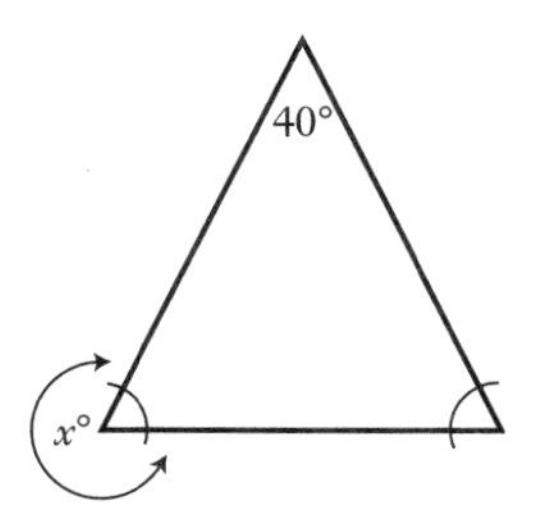

a. 70°
b. 110°
c. 140°
d. 290°

26. If the volume of a cube is 8 cubic inches, what is its surface area?

a. 80 square inches
b. 40 square inches
c. 24 square inches
d. 16 square inches

27. Giorgio is making a box. He starts with a 10×7 rectangle, then cuts 2×2 squares out of each corner. To finish, he folds each side up to make the box. What is the box's volume?

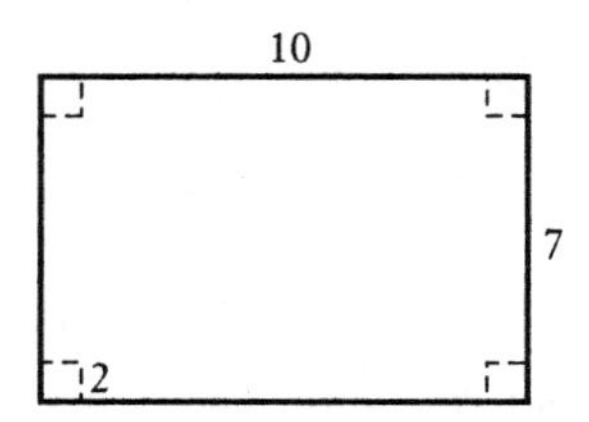

a. 36 squares
b. 42 squares
c. 70 squares
d. 72 squares

28. How many six-inch square tiles are needed to tile the floor in a room that is 12 feet by 15 feet?

a. 180 tiles
b. 225 tiles
c. 360 tiles
d. 720 tiles

Refer to the polygon below to answer questions 29 and 30:

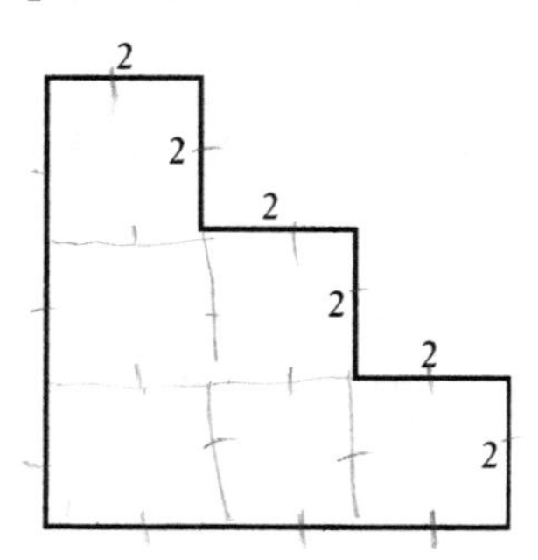

29. What is the perimeter of the polygon?

a. 8 units
b. 12 units
c. 20 units
d. 24 units

30. What is the area of the polygon?

- **a.** 8 square units
- **b.** 12 square units
- **c.** 20 square units
- **d.** 24 square units

31. The standard distance of a marathon is 26.2 miles. If the length of a walker's stride is 1.96 feet, approximately how many steps does she take to walk a marathon?

- **a.** 23,527
- **b.** 70,580
- **c.** 138,336
- **d.** 271,139

32. What is the measure of angle C in the following triangle?

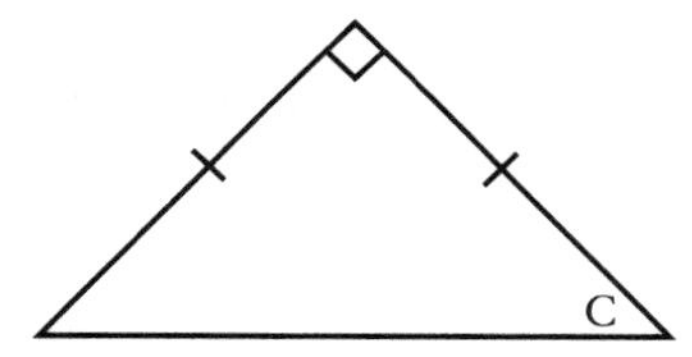

- **a.** 90°
- **b.** 60°
- **c.** 45°
- **d.** 25°

33. How much greater is the area of circle B?

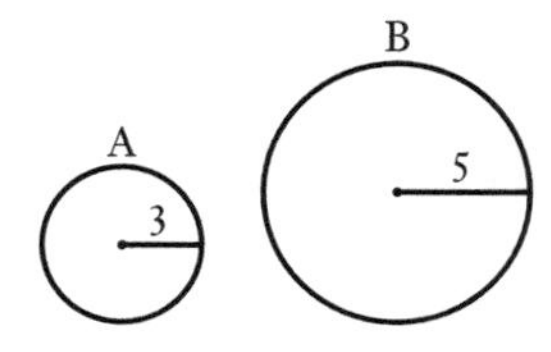

- **a.** 16π square units
- **b.** 9π square units
- **c.** 25π square units
- **d.** 14π square units

34.

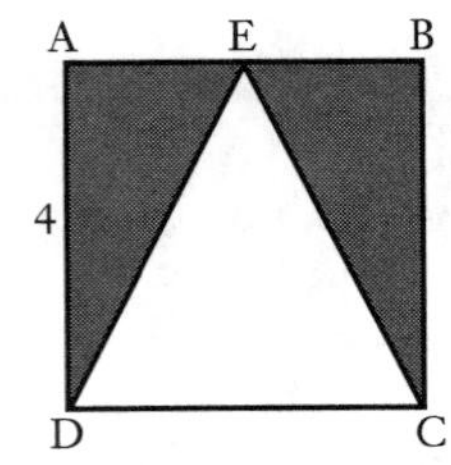

ABCD is a square and E is the midpoint of $\overline{AB}$. Find the area of the shaded region.

a. 4 square units
b. 6 square units
c. 8 square units
d. 12 square units

35. Two angles in quadrilateral ABCD have their measures indicated. The other two angles show variable expressions. What is x?

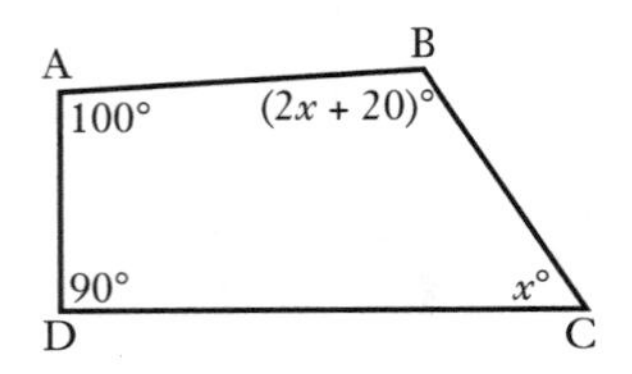

a. 50°
b. 60°
c. 70°
d. 80°

36. One cubic centimeter of clay weighs 3 grams. How much would a cube weigh if it measured 5 cm on each side?

a. 15 grams
b. 125 grams
c. 375 grams
d. 75 grams

Use the information and diagram below to answer questions 37 through 39:

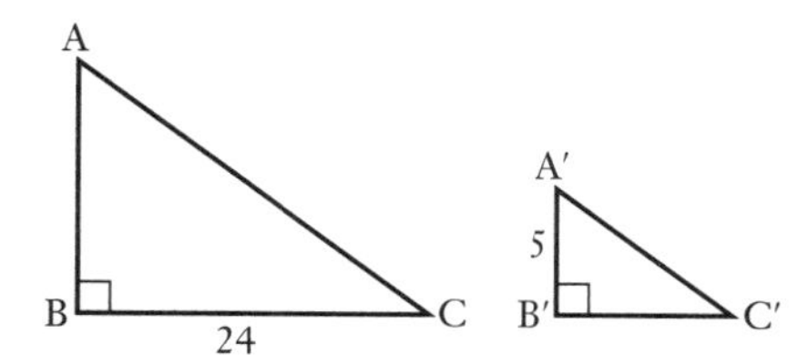

Note: All of the sides of ΔA′B′C′ are half the value of the sides of ΔABC.

37. Calculate the length of side A′C′ in triangle ΔA′B′C′.

a. 10
b. 12
c. 13
d. 26

38. The perimeter of ΔABC is how much greater than the perimeter of ΔA′B′C′?

a. 30
b. 40
c. 45
d. 60

39. The area of ΔABC is how much greater than the area of ΔA′B′C′?

a. 30
b. 40
c. 60
d. 90

40. What is the value of X in the figure below?

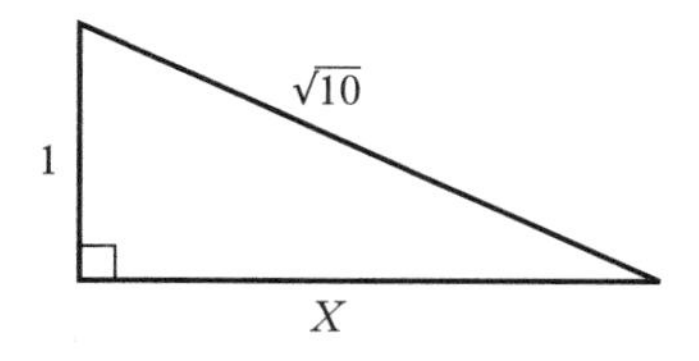

a. 3
b. 4
c. 5
d. 6

41. Find the area of the shaded portion in the figure below.

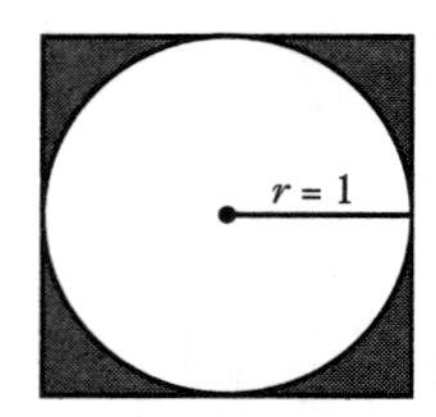

a. π
b. $\pi - 1$
c. $2 - \pi$
d. $4 - \pi$

42. What is the area of the shaded part of the circle below if the diameter is 6 inches? (Use 3.14 for π.)

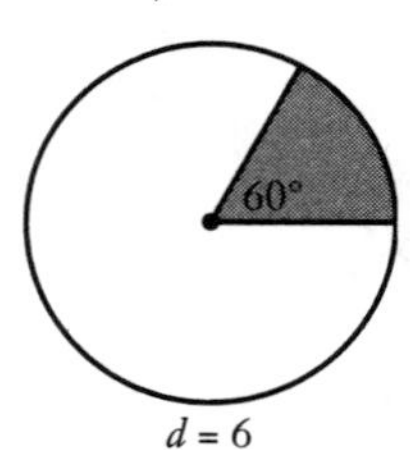

a. 4.71 square inches
b. 28.26 square inches
c. 60 square inches
d. 36 square inches

43. A cylindrical can measures 4.2 inches in height. Its circular bases of $\frac{1}{2}$ inch radii are removed, and the cylinder flattened out. What is the surface area of the flattened-out cylinder? (Use 3.14 for π)

a. 3.297 square inches
b. 8.54 square inches
c. 12.1 square inches
d. 13.188 square inches

44. A point on the outer edge of a wheel is 2.5 feet from the axis of rotation. If the wheel spins at a full rate of 2,640 revolutions per minute, how many miles will the point on the outer edge of the wheel travel in one hour?

a. 75π
b. 100π
c. 112π
d. 150π

45. What is the perimeter of the shaded area if the shape is a quarter-circle with a radius of 3.5? (Use $\pi = \frac{22}{7}$)

a. 7 units
b. 11 units
c. 22 units
d. 29 units

46. In the diagram, a half-circle is laid adjacent to a triangle. What is the total area of the shape, if the radius of the half-circle is 3 and the height of the triangle is 4?

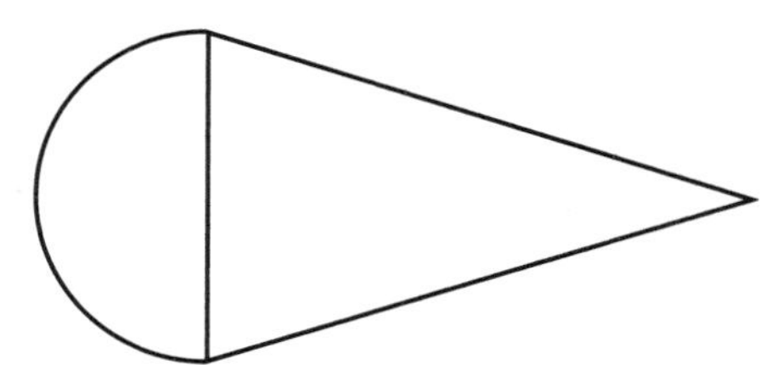

a. $6(\pi + 4)$
b. $6\pi + 12$
c. $6\pi + 24$
d. $\frac{9\pi}{2} + 12$

47. What is the area of the following shaded triangle?

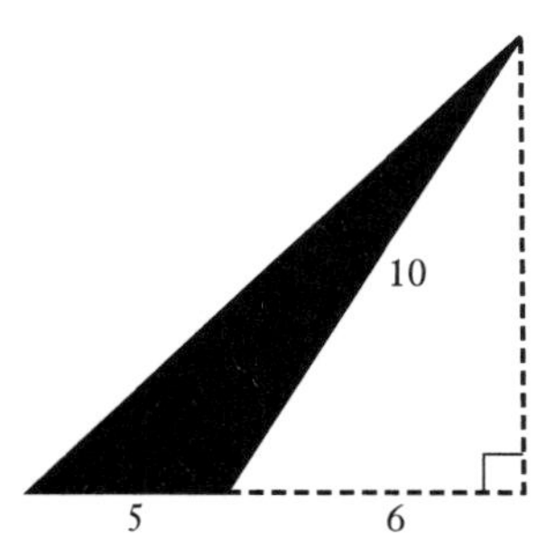

a. 20 square units
b. 25 square units
c. 40 square units
d. 44 square units

48. A triangle has sides that are consecutive even integers. The perimeter of the triangle is 24 inches. What is the length of the shortest side?

a. 10 inches
b. 8 inches
c. 6 inches
d. 4 inches

49. In the following diagram, a circle of area 100π square inches is inscribed in a square. What is the length of $\overline{AB}$?

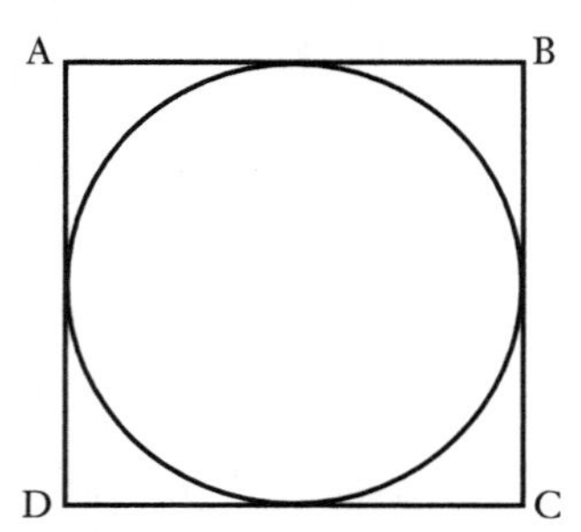

a. 10 inches
b. 20 inches
c. 40 inches
d. 100 inches

50. A bike wheel has a radius of 12 inches. How many revolutions will it take to cover 1 mile? (Use 1 mile = 5,280 feet, and $\pi = \frac{22}{7}$.)

a. 70
b. 84
c. 120
d. 840

ANSWERS

1. **d.** First, add up all of the given values:

3 ft. 5 in.
10 ft. 2 in.
+ 2 ft. 7 in.
15 ft. 14 in.

Next, note that 14 in. = 1 ft. + 2 in. This means 15 ft. 14 in. = 16 ft. 2 in., choice **d**.

2. **a.** First, add up all of the given values:

5 ft. 8 in.
4 ft. 7 in.
+ 3 ft. 9 in.
12 ft. 24 in.

Next, note that 24 in. = 2 ft., so 12 ft. 24 in. is equivalent to 14 ft.

3. **b.** Since there are 36 inches per yard, use the conversion factor $\frac{36 \text{ in.}}{1 \text{ yd.}}$, and multiply: $3\frac{1}{3}$ yd. × $\frac{36 \text{ in.}}{1 \text{ yd.}} = \frac{10}{3}$ yd. × $\frac{36 \text{ in.}}{1 \text{ yd.}} = \frac{360}{3}$ in. = 120 in.

4. **b.** 1 L = 1000 mL so you can use the conversion factor $\frac{1 \text{ L}}{1{,}000 \text{ mL}}$ to convert the milliliters into liters. 76,000 mL × $\frac{1 \text{ L}}{1{,}000 \text{ mL}}$ = 76 L.

5. **c.** Since there are 36 inches per yard, use the conversion factor $\frac{1 \text{ yd.}}{36 \text{ in.}}$ and multiply:

2,808 in. × $\frac{1 \text{ yd.}}{36 \text{ in.}}$ = 78 yd.

6. **c.** First, note that 4 yd. 6 in. is the same as 4 yd. $\frac{1}{2}$ ft., as this will help you combine units. Next, add up all the values:

5 yd. 2 ft.
8 yd. 1 ft.
3 yd. $\frac{1}{2}$ ft.
+ 4 yd. $\frac{1}{2}$ ft.
20 yd. 4 ft.

Next, note that 4 ft. = 1 yd. + 1 ft. Thus, 20 yd. 4 ft can be converted to 21 yd. 1 ft.

7. **a.** 1 mile equals 5,280 feet (memorize this). Since there are 3 feet per yard, use the conversion factor $\frac{1 \text{ yd.}}{3 \text{ ft.}}$ and multiply: 5,280 ft. × $\frac{1 \text{ yd.}}{3 \text{ ft.}}$ = 1,760 yd.

8. **a.** First, convert 3 ft. 5 in. into 36 in. + 5 in. = 41 in. Next, use the information given in the chart to make a conversion factor. Since 1 in. = 2.54 cm., and you want to end up with cm, you make a conversion factor with inches in the denominator: $\frac{2.54 \text{cm.}}{1 \text{ in.}}$. Next, multiply: 41 in. × $\frac{2.54 \text{cm.}}{1 \text{ in.}}$ = 104.14 cm.

9. **d.** The chart says that 1 yd. = .9 m., so you can write the conversion factor as $\frac{.9 \text{ m.}}{1 \text{ yd.}}$ and multiply: 5,500 yd. × $\frac{.9 \text{ m.}}{1 \text{ yd.}}$ = 4,950 m.

10. **c.** The chart says that 1 mi. = 1.6 km., so you can write the conversion factor as $\frac{1.6 \text{ km.}}{1 \text{ mi.}}$ and multiply: 1,280 mi. × $\frac{1.6 \text{ km.}}{1 \text{ mi.}}$ = 2,048 km.

11. **d.** Substitute 40 in for C in the given equation. Thus, ($F = \frac{9}{5}C + 32$) becomes $F = \frac{9}{5}(40) + 32 = (9)(8) + 32 = 72 + 32 = 104$ degrees Fahrenheit.

12. **a.** Line up the units and add:

45 min.
+ 1 hr. 25 min.
1 hr. 70 min.

Next, note that 70 min. = 1 hr. 10 min. Thus, 1 hr. 70 min. = 2 hr. 10 min.

13. **b.** First, convert the 12 yards into feet: 12 yd. $\times \frac{3 \text{ ft.}}{1 \text{ yd.}} = 36$ feet at the start. Next, Danielle cuts 2 feet off, so 34 feet are left.

14. **b.** Using the chart, you can make conversion factors where you will cross-off *pints* and end up with *ounces* (oz.). Thus, you multiply: 2 pt. $\times \frac{2 \text{ c.}}{1 \text{ pt.}} \times \frac{8 \text{ oz.}}{1 \text{ c.}} = 32$ oz.

15. **d.** Using the chart, you can make conversion factors where you will cross-off *ounces* and end up with *quarts* (qt.): 364 oz. $\times \frac{1 \text{ c.}}{8 \text{ oz.}} \times \frac{1 \text{ pt.}}{2 \text{ c.}} \times \frac{1 \text{ qt.}}{2 \text{ pt.}} = \frac{364}{32} = 11.375$ qt.

16. **a.** Using the chart, you can make conversion factors where you will cross-off *gallons* and end up with *ounces* (oz.): 3 gal. $\times \frac{4 \text{ qt.}}{1 \text{ gal.}} \times \frac{2 \text{ pt.}}{1 \text{ qt.}} \times \frac{2 \text{ c.}}{1 \text{ pt.}} \times \frac{8 \text{ oz.}}{1 \text{ c.}} = 384$ oz.

17. **c.** First, convert the gallons into quarts: 25 gal. $\times \frac{4 \text{ qt.}}{1 \text{ gal.}} = 100$ qt. If the fluid will fill 100 one-quart containers, it will then fill 200 $\frac{1}{2}$-quart containers.

18. **b.** If you draw a line on the diagram to denote the 45° angle mentioned, you can see that the angle section c makes with wall 2 must also be 45°. Recall that opposite angles formed by the intersection of two straight lines are equal:

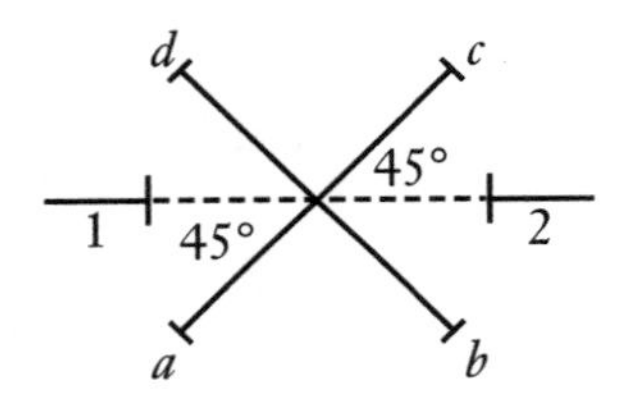

This means that section c makes a 45° angle with wall 2.

19. **c.** First, convert the width (1 yd.) into feet: 1 yd. = 3 ft. Next, use $A = lw = 6 \times 3 = 18$ ft^2. (Note that all of the answer choices are in ft^2, so converting to feet is a good idea.)

20. **a.** The area of the square is $A = s^2 = 6^2 = 36$ square cm. The area of the rectangle must then also be 36 cm^2. Substituting this into the area formula, along with $l = 9$ we get: $A = lw$; $36 = 9 \times w$; $w = 36 \div 9 = 4$ cm.

21. **d.** You are told that Area = 9π. If A = πr^2, then $\pi r^2 = 9\pi$, and $r = 3$. Circumference, $C = 2\pi r = 2\pi \times 3 = 6\pi$ cm. Remember that perimeters and circumferences are measured in units (like cm.) and areas are measured in square units (like cm^2).

22. **b.** First, calculate the area in square feet. The area of a rectangle is lw, so $A = lw = 860$ ft. $\times$ 560 ft. = 481,600 ft^2. Next, use the conversion factor $\frac{1 \text{ acre}}{43{,}560 \text{ ft}^2}$ and multiply: 481,600 ft^2 $\times \frac{1 \text{ acre}}{43{,}560 \text{ ft}^2}$ $\approx$ 11.056 acres $\approx$ 11.06 acres.

23. **c.** Area = lw. First, convert the mixed numbers to improper fractions: $1\frac{2}{7}$ in. = $\frac{9}{7}$ in. and $1\frac{4}{5}$ in. = $\frac{9}{7}$ in. Next, use these fractions in the formula: Area = $lw = \frac{9}{7} \times \frac{9}{5} = \frac{81}{35}$ in.^{2} = $2\frac{11}{35}$ square inches.

24. **b.** The perimeter of a rectangle is the sum of all its sides: 160 + 160 + 80 + 80 = 480 feet. Next, convert to yards by multiplying 480 with the conversion factor $\frac{1 \text{ yd.}}{3 \text{ ft.}}$: 480 ft. × $\frac{1 \text{ yd.}}{3 \text{ ft.}}$ = 160 yd.

25. **d.** The curved markings indicate that the 2 bottom angles are equal. We can call these 2 equal angles y. Thus, $y + y + 40° = 180°$, $2y + 40° = 180°$; $2y = 140°$; $y = 70°$. Angles x and y form a complete circle (360°). Thus, $x = 360° - y° = 360° - 70° = 290°$.

26. **c.** The volume formula for a cube is $V = s^3$, so here $s^3 = 8$ and $s = 2$ in. The surface area of one face is $s^2 = 2^2 = 4$ square inches. Since there are six faces, the total surface area is 6 × 4 square inches = 24 square inches.

27. **a.** When the 2 × 2 squares are cut out, the length of the box is 3, and the width is 6. The height is 2:

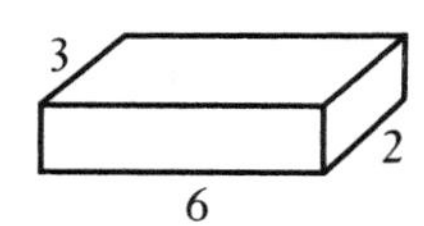

The volume is 3 × 6 × 2, or 36.

28. **d.** Draw yourself a rectangle to represent the 12 ft. × 15 ft. floor. Since each tile is 6 in. by 6 in., or $\frac{1}{2}$ **ft. by** $\frac{1}{2}$ **ft.**, you can see that you could get 24 tiles across the floor, and 30 tiles going down. Now, you just multiply 24 by 30 to get the total tiles needed: 24 × 30 = 720.

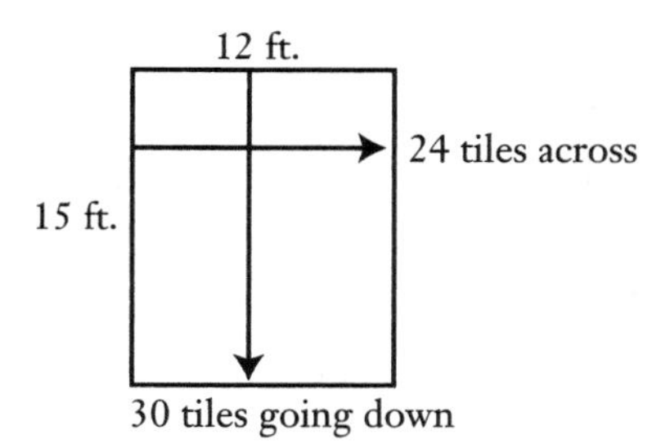

29. **d.** Fill in the missing sides:

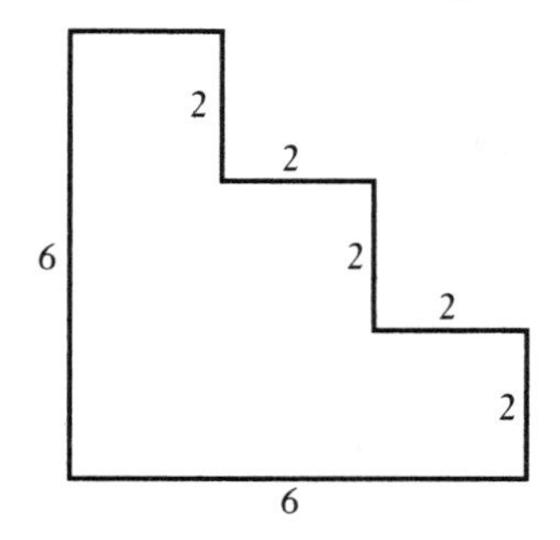

Next, add up all the sides: $P = 6 + 6 + 6(2) = 12 + 12 = 24$ units.

30. **d.** Divide up the figure into squares as shown below:

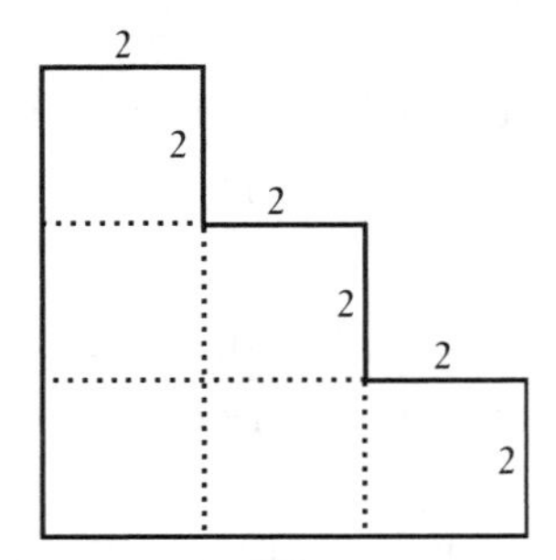

The figure is composed of 6 squares. The area of each square is $s^2 = 2^2 = 4$. Thus the total area is $6 \times 4 = 24$ square units.

31. **b.** Convert 26.2 miles to feet, and divide by the length of the walker's stride to find how many steps she takes in a marathon: 1 mile = 5,280 feet, so 26.2 miles = 138,336 feet. Divide 138,336 by 1.96 feet per step to get 70,579.6. Round to the nearest whole number to get 70,580 steps.

32. **c.** The two lines through the sides of the triangle indicate that they are equal. The right angle is 90° and the 2 angles opposite the 2 equal sides will be equal. Since the interior angles of a triangle add to 180°, the 2 equal angles must add to 180° – 90° = 90°. Thus, each angle will be equal to 45°. Thus, angle $C = 45°$.

33. **a.** Remember the formula for figuring out the area of a circle: $A = \pi r^2$. Circle A then is $\pi 3^2$ or 9π and circle B is $\pi 5^2$ or 25π, so the area of circle B is 16π greater than circle A.

34. **c.** To find the area of the shaded region, simply subtract the area of the triangle from the area of the square. The area of the triangle is $\frac{1}{2}bh = \frac{1}{2}(4)(4) = 8$ square units, and the area of the square is $s^2 = 4^2 = 16$ square units. Thus, the area of the shaded region is 16 – 8 = 8 square units.

35. **a.** Set up an equation. (Remember, all the angles added up inside a four-sided figure equals 360°): $90 + 100 + x + 2x + 20 = 360$, which is $3x + 210 = 360$. Subtract 210 from both sides to get $3x = 150$. Divide by 3 to get $x = 50$.

36. **c.** For this question, you already know that the weight is $\frac{3\text{ g}}{\text{cm}^3}$. You need to find out how many cm^3 there are in the given cube, which is the *volume* of the cube. For a cube, the *volume* = *side*^{3}. The given cube has a side = 5, so $V = 5^3 = 5 \times 5 \times 5 = 125$. Then, to find the weight you multiply $125\text{ cm}^3 \times \frac{3\text{ g}}{\text{cm}^3} = 375$ grams for the answer.

37. **c.** Since $\overline{BC} = 24$, B′C′ will be half that, or 12. Thus, ΔA′B′C′ is a right triangle with legs equaling 5 and 12. You can use the Pythagorean theorem to solve for the hypotenuse: $a^2 + b^2 = c^2$ becomes $5^2 + 12^2 = c^2$, then $25 + 144 = c^2$, then $169 = c^2$, so c = 13.

38. **a.** ΔA′B′C′ is a 5-12-13 right triangle (see answer explanation for question 37) and ΔABC is double that, or 10-24-26. Thus, the perimeter of ΔA′B′C′ is 5 + 12 + 13= 30, and the perimeter of ΔABC is twice that, or 60. Thus, the difference is 60 – 30 = 30.

39. **d.** ΔA′B′C′ is a 5-12-13 right triangle (see answer explanation for question 37) and ΔABC is double that, or 10-24-26. The base of ΔA′B′C′ is 24, and its height is 10. Apply the area formula: $A = \frac{1}{2}bh = \frac{1}{2}(24)(10) = 120$ units². The base of ΔABC is 12, and its height is 5. Apply the area formula: $A = \frac{1}{2}bh = \frac{1}{2}(12)(5) = 30$ units². Thus, the difference is 120 – 30 = 90 units.

40. **a.** You can use the Pythagorean theorem to solve for the missing leg: $a^2 + b^2 = c^2$ becomes $1^2 + X^2 = (\sqrt{10})^2$, then $1 + X^2 = 10$, so $X^2 = 9$, and $X = 3$.

41. **d.** The shaded area is the difference between the area of the square and the circle. Because the radius is 1, a side of the square is 2. The area of the square is $s^2 = 2^2 = 4$, and the area of the circle is $\pi r^2 = \pi 1^2 = \pi$. Therefore, the answer is $4 - \pi$.

42. **a.** First, find the area of the circle: Area = πr^2, or 3.14×9, which equals 28.26 square inches. Then, notice there are 360° in a circle and 60° is one-sixth that ($\frac{60}{360} = \frac{1}{6}$). The shaded area is then only one-sixth the area of the total circle. So, you simply divide 28.26 by 6 to get 4.71 square inches.

43. **d.** After removing the circular bases, you are left with a flat rectangle. Since the height was 4.2 in, the length of the rectangle is 4.2 in. Since the circumference of the bases was $C = 2\pi r = 2 \times 3.14 \times \frac{1}{2} = 3.14$ in., the width of the rectangle is 3.14 in. Thus, the area of the new rectangular figure is $lw = 4.2 \times 3.14 = 13.188$ in^2.

44. **d.** The point lies on the circumference of a circle with a radius of 2.5 feet. Therefore, the distance that the point travels in one rotation is the length of the circumference of the circle, or $2\pi r = 2\pi(2.5) = 5\pi$ feet. Since the wheel spins at 2,640 revolutions per minute, the point travels $2{,}640 \times 5\pi$ feet per minute = $13{,}200\pi$ feet per minute. Multiplying by 60 to find the distance traveled in one hour, you get $60 \times 13{,}200\pi = 792{,}000\pi$ feet per hour. Dividing by 5,280 feet to convert to miles, you get 150π miles per hour.

45. **d.** The curved length of the perimeter is one quarter of the circumference of a full circle: $\frac{1}{4}2\pi r$, $= 2(\frac{22}{7})(3.5) = 7 \times \frac{22}{7} = 22$. The linear (straight) lengths are radii, so the solution is simply 22 + 2(3.5); or 29.

46. **d.** Because the radius of the hemisphere is 3, and it is the same as half the base of the triangle, the base must be 6. Therefore, the area of the triangle is $\frac{1}{2}bh = \frac{1}{2}(4 \times 6) = 12$. The area of the circle, if it was a whole circle, is πr^2, which equals 9π. Therefore, the area of a half-circle is $\frac{9\pi}{2}$. Adding gives $\frac{9\pi}{2} + 12$.

47. **a.** To get the height of the triangle (h), using the Pythagorean theorem: $a^2 + b^2 = c^2$ becomes $6^2 + h^2 = 10^2$, then $36 + h^2 = 100$, and $h^2 = 64$, so the height, h, equals 8. Then, 5 is plugged in for the base and 8 for the height in the area equation $A = \frac{1}{2}bh$. Thus, $A = \frac{1}{2}(5)(8) = 20$ square units.

48. **c.** An algebraic equation can be used to solve this problem. The shortest side can be denoted s. Therefore, $s + (s + 2) + (s + 4) = 24$; $3s + 6 = 24$, and $s = 6$.

49. **b.** If the circle is 100π square inches, its radius must be 10 inches (because $A = \pi r^2$ and here $A = 100\pi$). $\overline{AB}$ is twice the radius, so it is 20 inches.

50. **d.** The outer edge of the wheel is in contact with the ground. Since you are told to use 1 mile = 5,280 feet, you would be wise to convert the 12 in. radius to 1 ft. You can find the outer edge (circumference) by using $C = 2\pi r = 2(\frac{22}{7})(1) = \frac{44}{7}$ ft. Thus, each time it revolves it covers $\frac{44}{7}$ ft. Divide 5,280 feet by $\frac{44}{7}$ feet to find the number of revolutions in 1 mile: $5{,}280 \div \frac{44}{7} = 5{,}280 \times \frac{7}{44} = 840$ revolutions.

Practice Test 1

This chapter contains your first practice test. After reviewing the chapters in this book you should be able to put all that you have learned together and take these sample examinations. Take Practice Test 1. Be sure to re-evaluate the questions you answered incorrectly by going back and studying the necessary material from earlier chapters. Then try it again: Take Practice Test 2 in the next chapter. Each test should take one hour to complete. Good luck!

1. If a piece of packaging foam is .05 in thick, how thick would a stack of 350 such pieces of foam be?

a. 7,000 in.

b. 700 in.

c. 175 in.

d. 17.5 in.

2. 30% of what number equals 60% of 9,000?

a. 18,000

b. 5,400

c. 2,400

d. 1,620

3. Three pieces of wood measure 4 yd. 1 ft. 3 in., 5 yd. 2 ft. 4 in., and 4 yd. 1 ft. 5 in. lengthwise. When these boards are laid end to end, what is their combined length?

a. 14 yd. 2 ft.

b. 14 yd. 1 ft. 11 in.

c. 13 yd. 2 in.

d. 13 yd. 2 ft.

4. Select the answer choice that best completes the sequence below.

a.

b.

c.

d.

5. During a race, markers will be placed along a roadway at regular .2-mile intervals. If the entire roadway is 10,560 feet long, how many such markers will be used?

a. 10

b. 100

c. 20

d. 200

6. If it takes 27 nails to build 3 boxes, how many nails will it take to build 7 boxes?

a. 64

b. 72

c. 56

d. 63

7. The average purchase price (arithmetic mean) of four shirts is \$9. If one shirt was priced at \$15, and another at \$7, what might be the prices of the other 2 shirts?

a. \$4 and \$3
b. \$7 and \$15
c. \$9 and \$9
d. \$10 and \$4

8. What percent of $\frac{3}{8}$ is $\frac{1}{2}$?

a. 25%
b. $33\frac{1}{3}$%
c. 75%
d. $133\frac{1}{3}$%

9. A large bag of cement mix weighs $38\frac{1}{2}$ pounds. How many quarter-pound bags of mix can be made from this large bag?

a. under 10 bags
b. 16 bags
c. 80 bags
d. 154 bags

10. Use ($F = \frac{9}{5}C + 32$) to convert 15° C into the equivalent Fahrenheit temperature.

a. 59°
b. 60°
c. 62°
d. 65°

11. What is the perimeter of the shaded area if the shape is a quarter circle with a radius of 8?

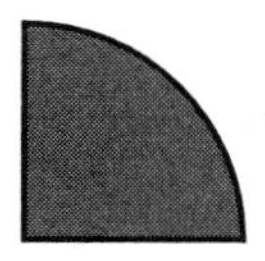

a. 2π
b. 4π
c. $2\pi + 8$
d. $4\pi + 16$

12. Select the answer choice that best completes the sequence below.
CMM EOO GQQ ______ KUU

a. GRR
b. GSS
c. ISS
d. ITT

13. How many ounces are in 5 pints?

a. 10 oz.
b. 20 oz.
c. 40 oz.
d. 80 oz.

14. A rod that is 3.5×10^7 cm. is how much shorter than a rod that is 7×10^{14} cm.?

a. 20,000,000 times shorter
b. 4,000,000 times shorter
c. 50,000 times shorter
d. 20,000 times shorter

15. Joel had to insert form letters into 800 envelopes. In the first hour, he completed $\frac{1}{8}$ of the total. In the second hour, he completed $\frac{2}{7}$ of the remainder. How many envelopes does he still have to fill?

a. 300
b. 400
c. 500
d. 700

16. Jen's median bowling score is greater than her mean bowling score for five tournament games. If the scores of the first four games were 140, 192, 163, and 208, which could have been the score of her fifth game?

a. 130
b. 145
c. 168
d. 177

17. An 18-gallon barrel of liquid will be poured into containers that each hold half of a pint of fluid. If all of the containers are filled to capacity, how many will be filled?

a. 36
b. 72
c. 144
d. 288

18. Select the answer choice that best completes the sequence below.

Ô Ö ∅ ⦸ | O̬ O̤ ⦸ ∅ | Ô ___ ___ ⦸

a. ∅ O̬

b. Ö ⦸

c. ⦸ Ö

d. Ö ∅

19. In a box of 300 nails, 27 are defective. If a nail is chosen at random, what is the probability that it will not be defective?

a. $\frac{27}{100}$

b. $\frac{91}{100}$

c. $\frac{27}{300}$

d. $\frac{91}{300}$

20. When Christian and Henrico work together they can complete a task in 6 hours. When Christian works alone he can complete the same task in 10 hours. How long would it take for Henrico to complete the task alone?

a. 45

b. 30

c. 15

d. 10

21. The square root of 52 is between which two numbers?

a. 6 and 7

b. 7 and 8

c. 8 and 9

d. none of the above

22. Juliet made \$12,000 and put $\frac{3}{4}$ of that amount into an account that earned yearly interest at a rate of 4%. After 3 years, what is the dollar amount of the interest earned?

a. \$10,080

b. \$10,800

c. \$1,800

d. \$1,080

23. If the area of a circle is 16π square inches, what is the circumference?

a. 2π inches

b. 4π inches

c. 8π inches

d. 12π inches

24. Select the answer choice that best completes the sequence below.

QAR RAS SAT TAU _____

a. UAV
b. UAT
c. TAS
d. TAT

25. A container was filled $\frac{1}{3}$ of the way with fluid. Damian added 24 liters more, filling the container to full capacity. How many liters are in the container now?

a. 12 L
b. 30 L
c. 36 L
d. 48 L

26. Bolts cost $4 per 10 dozen and will be sold for 10¢ each. What is the rate of profit?

a. 200%
b. 150%
c. 100%
d. 75%

27. Select the answer choice that best completes the sequence below.

____

a.
b.
c.
d.

28. $6,000 is deposited into an account. If interest is compounded semiannually at 2% for 6 months, then what is the new amount of money in the account?

a. $120
b. $6,060
c. $240
d. $6,240

A forest fire engulfed the Wildlife Preserve in Blackhill County in 1998. Since then, park rangers have kept track of the number of forest animals living in the forest. Below is a graph of how many deer, foxes, and owls were reported during the years following the fire. Use this information to answer questions 29 through 32.

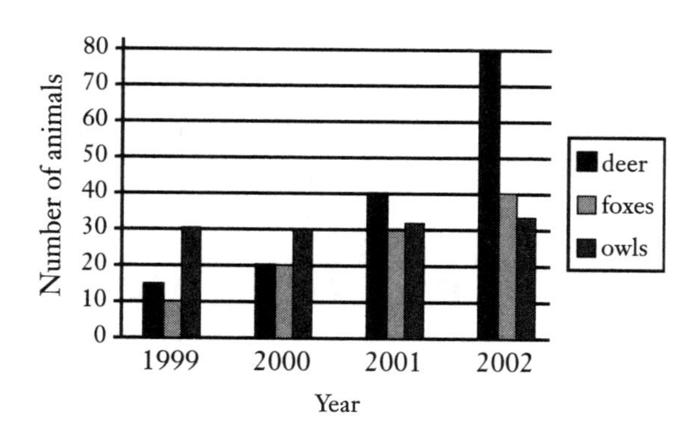

29. Which of the following statements appears to be true for the years shown?

a. The fox population doubled every year since 1999.
b. The deer population doubled every year since 2000.
c. The owl population showed neither a steady increase nor decrease.
d. Both **b** and **c** are true.

30. Which statement might explain the data presented in the graph?

a. The owl population was greatly reduced by the fire, and, thus, the trend shows a steady increase in this population during the years of recovery.
b. The owls were able to fly away from the fire, thus the owl population does not show the pattern of recovery that the deer and fox population exhibit.
c. Factors independent of the fire are causing a steady decline in the owl population.
d. A steep decline in the owl population can be attributed to illness.

31. The growth of the deer population from 2001–2002 was how much greater than the growth of the fox population for the same year?

a. 10
b. 20
c. 30
d. 40

32. What was the percent increase in deer from 1999–2000?

a. $33\frac{1}{3}\%$
b. 50%
c. $\frac{3}{4}\%$
d. $\frac{1}{3}\%$

33. A square with *sides* = 8 in. has the same area of a rectangle with *width* = 4 in. What is the length of the rectangle?

a. 8 in.
b. 12 in.
c. 16 in.
d. 64 in.

34. A rectangular tract of land measures 440 feet by 1,782 feet. What is the area in acres? (1 acre = 43,560 square feet.)

a. 14 acres
b. 16 acres
c. 18 acres
d. 20 acres

35. What is the mode of the following numbers?
12, 9, 8, 7, 8, 9, 5, 9

a. 7
b. 8.375
c. 9
d. 9.5

36. The largest sector of the pie chart below has a central angle equal to how many degrees?

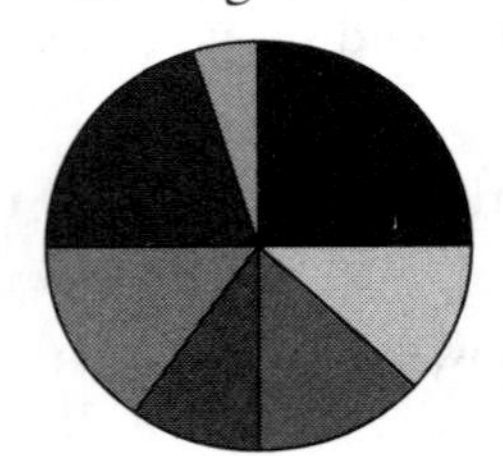

a. 15 degrees
b. 45 degrees
c. 90 degrees
d. 180 degrees

37. The chart below shows the monthly attendance for union meetings over the course of 4 months. Which 2 months had the same number of members attending?

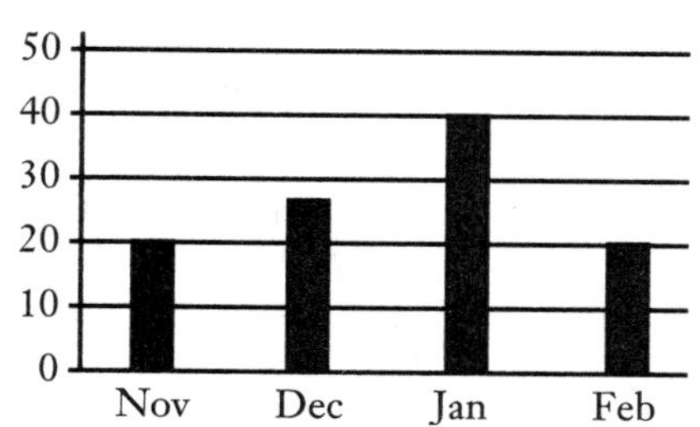

a. November and December
b. December and February
c. November and February
d. December and January

38. If the radius of a cylindrical tank is 7 cm. and its volume is 1,540 cm^3, what is the height in cm?
a. 10 cm.
b. 15.4 cm.
c. 10π cm.
d. 15.4π cm.

39. If Martin exchanges 120 quarters, 300 dimes, 600 nickels, and 500 pennies for bills, he may get
a. 4 twenty-dollar bills, 2 ten-dollar bills, and 1 five-dollar bill.
b. 3 twenty-dollar bills, 1 ten-dollar bill, and 1 five-dollar bill.
c. 2 fifty-dollar bills and 1 twenty-dollar bill.
d. 1 fifty-dollar bills, 2 twenty-dollar bills, and 1 five-dollar bill.

40. Brian jogged 12 miles. For the first 2 miles, his pace was 3 mph. For the next 3 miles, his pace was 5 mph. For the remainder of his jog, his pace was 4 mph. What was his average speed?
a. 4.2 mph
b. 6.86 mph
c. 7.2 mph
d. $2\frac{2}{3}$ mph

ANSWERS

1. d. To solve, simply multiply the thickness of each piece of foam by the total number of pieces. $.05 \times 350 = 17.5$ in.

2. a. "30% of what number equals 60% of 9,000?" can be written mathematically as $.30 \times x = .60 \times 9{,}000$. Dividing both sides by .30 will yield

$$x = \frac{(.60)(9{,}000)}{.30} = \frac{5{,}400}{.30} = 18{,}000.$$

3. a. First, line up all of the units and add:

4 yd. 1 ft. 3 in.
5 yd. 2 ft. 4 in.
+ 4 yd. 1 ft. 5 in.
13 yd. 4 ft. 12 in.

Next, note that 12 in. = 1 ft., so 13 yd. 4 ft. 12 in. is the same as 13 yd. 5 ft., and that 3 ft. = 1 yd., so 5 ft. = 1 yd. + 2 ft. Ultimately, you can rewrite the entire length as 14 yd. 2 ft.

4. d. The amount of the shaded area changes from $\frac{1}{2} \rightarrow \frac{1}{4} \rightarrow \frac{1}{2}$. Thus, you need to find the answer that is $\frac{1}{4}$ shaded, followed by $\frac{1}{2}$ shaded. Choice **d** is correct.

5. a. 5,280 feet = 1 mile, so 10,560 feet = 2 miles. To solve, divide the total 2 mile distance by the interval, .2 miles: $2 \div .2 = 10$.

6. d. First, set up a proportion: $\frac{27}{3} = \frac{x}{7}$. You can reduce the first fraction: $\frac{9}{1} = \frac{x}{7}$ and then cross-multiply: $1(x) = 9(7)$, so $x = 63$.

7. d. If the cost of 4 shirts averaged out to \$9, then the sum of all four shirts was $4 \times 9 = \$36$. (Note that the sum of all 4 shirts must equal \$36 in order for the average to equal 9: Average = sum $\div$ 4 = 36 $\div$ 4 = 9.) Of the \$36 total, \$22 is accounted for (one shirt was \$15, and another \$7), leaving \$14 unaccounted for. Only choice **d** adds to \$14.

8. d. Recall that "What percent" can be expressed as $\frac{x}{100}$. The question, "What percent of $\frac{3}{8}$ is $\frac{1}{2}$?" can be expressed as: $\frac{x}{100} \cdot \frac{3}{8} = \frac{1}{2}$. This simplifies to $3 \cdot \frac{x}{800} = \frac{1}{2}$. Cross-multiplying yields $6 \times x = 800$. Dividing both sides by 6 yields $x = 133\frac{1}{3}\%$.

9. d. Divide $38\frac{1}{2}$ by $\frac{1}{4}$. By expressing $38\frac{1}{2}$ as its equivalent 38.5, you get: $38.5 \div \frac{1}{4} = 38.5 \times \frac{4}{1} = 154$ bags.

10. a. Substitute 15° C in for the variable C in the given equation. Thus, ($F = \frac{9}{5}C + 32$) becomes $F = \frac{9}{5}(15) + 32 = (9)(3) + 32 = 27 + 32 = 59$ degrees Fahrenheit.

11. d. The perimeter of the curved length is a quarter of the circumference of a whole circle when $r = 8$. Since $C = 2\pi r$ and you want a quarter of this value, solve $\frac{1}{4} \times 2 \times \pi \times r = \frac{1}{4} \times 2 \times \pi \times 8 = 4\pi$. The 2 straight edges are radii and are each 8 units long. Thus, the total perimeter $= 4\pi + 8 + 8 = 4\pi + 16$.

12. **c.** The first letter of each triplet changes by skipping 1 letter : C → E → G → **I** →K. Thus, the first letter in the missing triplet is I. The last 2 letters of each triplet follow the same pattern (skip 1 letter): MM → OO → QQ → **SS** → UU. Thus, the answer is ISS.

13. **d.** Using the knowledge that 1 pt. = 2 c. and 1 c. = 8 oz., you can use a series of conversion factors to eliminate pints and keep ounces. Thus, you multiply: 5 pt. $\times \frac{2\text{ c.}}{1\text{ pt.}} \times \frac{8\text{ oz.}}{1\text{ c.}} = 80$ oz.

14. **a.** To find how many "times shorter" the first rod is, just divide: $\frac{7 \times 10^{14}}{3.5 \times 10^{7}} = 2 \times 10^{14-7} = 2 \times 10^{7}$ = 20,000,000 times shorter.

Hint: Treat their division like two separate division operations, $7 \div 3.5$ and $10^{14} \div 10^{7}$. BUT, you must remember that the dividends are ultimately multiplied together in the end. Also, to divide 10^{14} by 10^{7}, simply subtract the exponents.

15. **c.** Joel starts with 800 envelopes to fill. During the first hour he filled $\frac{1}{8}$ of the 800: $\frac{1}{8} \times 800 = 100$. He then had 800 – 100 = 700 left to fill. In the second hour he filled $\frac{2}{7}$ of the remaining 700. $\frac{2}{7} \times 700 = 200$ filled in the second hour. After two hours, Joel has 700 – 200 = 500 remaining.

16. **d.** The mean is found by adding up the numbers and dividing by the number of values. The median is found by listing all of the numbers in order and taking the middle value. To find the solution, try out each answer choice to see if it works. A score of 130 would give a mean of 167 and a median of 163. A score of 145 would give a mean of 169 and a median of 163. A score of 168 would give a mean of 174 and a median of 168. A score of 177 would give a mean of 176 and a median of 177. 177 is the only one which has a median greater than the mean:

Median = 140 163 ***177*** 192 208

Mean = (140 + 163 + 177 + 192 + 208) ÷ 5 = 880 ÷ 5 = ***176***

17. **d.** Using the knowledge that 1 gal. = 4 qt. and 1 qt. = 2 pt., you can generate a series of conversion factors and multiply them so that you can cross out the units you do not want (gal.) and keep the units you do want: 18 gal. $\times \frac{4\text{ qt.}}{1\text{ gal.}} \times \frac{2\text{ pt.}}{1\text{ qt.}} = 144$ pints. Next, remember you are looking for $\frac{1}{2}$-pints. 144 pints will fill 288 half-pint containers.

18. **d.** This is an alternating series. The first and third segments are repeated. The second segment is simply a reverse of the other two.

19. **b.** If 27 of the 300 are defective, then 300 – 27 = 273 are not defective. Thus, the probability of selecting a nail that is not defective will be 273 out of 300:

$\frac{273}{300} = \frac{91}{100}$.

20. **c.** Christian can complete $\frac{1}{10}$ of the task in 1 hour. (You assume this because he completes the entire task in 10 hours.) Together, Christian and Henrico complete $\frac{1}{6}$ of the task in 1 hour. Convert both fractions into *thirtieths*. $\frac{5}{30}$ per hour (both men) – $\frac{3}{30}$ per hour (just Christian) = $\frac{2}{30}$ = $\frac{1}{15}$ per hour (just Henrico). Since Henrico completes $\frac{1}{15}$ of the task per hour, it will take him 15 hours to complete the entire task when working alone.

21. **b.** $7^2 = 49$ and $8^2 = 64$. So the square root of 52 will equal a number that is between 7 and 8.

22. **d.** Use the formula: $I = PRT$, which means *Interest = principal × rate of interest × time*. Where principal equals your original amount of money (in dollars), and time is in years. Here the original amount of money (P) is $9,000 because she put $\frac{3}{4}$ of the $12,000 into the account. $I = .04$ and $T = 3$ years. Substituting into $I = PRT$, you get $I = (9{,}000)(.04)(3) = \$1{,}080$.

23. **c.** You are told that Area = 16π . Since $A = \pi r^2$, $16 = r^2$, and $r = 4$. Use this r in the circumference formula: Circumference = $C = 2\pi r = 2\pi \times 4 = 8\pi$ inches.

24. **a.** The first letter in each triplet progresses from Q → R → S → T, so the next triplet will begin with U. The second letter of each triplet is a constant: A. The third letter of each triplet progresses from R → S → T → U, so the 3rd letter in the next triplet will be V. Thus, the answer is UAV.

25. **c.** 24 L represents $\frac{2}{3}$ of the whole capacity. You can ask yourself, "$\frac{2}{3}$ of what number is 24?" This can be expressed mathematically as $\frac{2}{3} \times x = 24$; $x = 24 \div \frac{2}{3} = 24 \times \frac{3}{2} = 36$ L.

26. **a.** 10 dozen bolts = $10 \times 12 = 120$ bolts. When they are all sold, the amount collected is $\$.10 \times 120 = \$12$. Since the 10 dozen cost $4, the profit is $\$12 - \$4 = \$8$. Next, to find the rate of profit, set up a proportion:

$$\frac{\$8 \text{ profit}}{\text{initial } \$4} = \frac{x}{100}.$$

Cross-multiply to get $(100)(8) = (4)(x)$, or $800 = (4)(x)$. Divide both sides by 4 to get $x = 200$. Thus, the rate of profit is 200%.

27. **a.** As the series progresses, the amount of shading changes from $\frac{1}{2}$ → $\frac{3}{4}$ → whole → none → $\frac{1}{4}$ → $\frac{1}{2}$ → $\frac{3}{4}$ → whole. So the next 2 terms will be: none → $\frac{1}{4}$.

28. **b.** Because the interest is compounded semiannually (twice a year), after $\frac{1}{2}$ a year (6 months) the amount of interest earned $I = PRT = 6{,}000 \times .02 \times \frac{1}{2} = \$60$. Now, the account has $\$6{,}000 + \$60 = \$6{,}060$ in it.

29. **d.** The fox population (lightest-colored bars) went up by 10 animals each year. Thus, choice **a** is wrong. The deer population (black bar) doubled every year since 2000 → 20 → 40 → 80. The owl population stayed around 30, showing neither an increase nor decrease. Thus, both **b** and **c** are true statements, making choice **d**: "Both **b** and **c** are true," the correct answer.

30. **b.** The owl population is maintaining a steady rate of growth. There is not a steady increase (**a** is wrong), a steady decline (**c** is wrong), or a steep decline (**d** is wrong). Thus, choice **b** is the correct answer.

31. **c.** The deer (black bar) went from 40 in 2001 to 80 in 2002. That is an increase of 40 deer. The fox population (lightest-colored bar) grew from 30 in 2001 to 40 in 2002. That is an increase of 10. Thus, the difference in growths is $40 - 10 = 30$.

32. **a.** The deer (black bar) increased from 15 in 1999 to 20 in 2000. This is a change of 5 deer. When compared to the initial 15, 5 out of 15 represents $\frac{5}{15} = \frac{x}{100}$; $x = 33\frac{1}{3}\%$.

33. **c.** The area of the square is $A = side^2 = s^2 = 8^2 = 64$ in^2. The area of the rectangle must then also be 64 in^2. Substituting this area and the given width $w = 4$ into the area formula, you get: $A = lw$; $64 = l \times 4$; $l = 64 \div 4 = 16$ in.

34. c. First, calculate the area in square feet: Area = lw = 440 ft × 1,782 ft = 784,080 ft^2. Next convert to acres by using the conversion factor $\frac{1 \text{ acre}}{43{,}560 \text{ ft}^2}$ and multiply: 784,080 ft^2 × $\frac{1 \text{ acre}}{43{,}560 \text{ ft}^2}$ = 18 acres.

35. c. The mode is the number that occurs the most. You are given:

12, **9**, 8, 7, 8, **9**, 5, **9**.

Note that 9 occurs the most and is the mode.

36. c. The largest sector takes up a quarter of the pie chart (the gray sector). The interior angles of a circle add to 360 degrees and $\frac{1}{4}$ of 360 = $\frac{1}{4}$ × 360 = 90 degrees.

37. c. The attendance for both November and February was 20 members each. You can tell that this is true because the bars for these months are the same height.

38. a. If you use $\pi = \frac{22}{7}$, and the formula $V = \pi r^2 h$, you get $1{,}540 = \frac{22}{7} \times 7^2 \times h$. This simplifies to $1{,}540 = 154 \times h$. Dividing both sides by 154 yields h = 10 cm.

39. d. Multiply the number of coins by the value of the coin:
120 quarters = 120 × \$.25 = \$30
300 dimes = 300 × \$.10 = \$30
600 nickels = 600 × \$.05 = \$30
500 pennies = 500 × \$.01 = \$5
Next, add all of the dollar amounts: \$30 + \$30 + \$30 + \$5 = \$95. The only choice that represents \$95 is **d**: 1 fifty-dollar bill, 2 twenty-dollar bills, and 1 five-dollar bill.

40. b. To find the average speed, you must use $D = RT$ (*Distance* = *Rate* × *Time*) with the total distance and the total time as D and T, *respectively*. You are given the total distance of 12 miles. You need the total time. This can be found by using the information in the question. The formula $D = RT$ can be rewritten as $T = \frac{D}{R}$. Making a chart for yourself will help you stay organized:

INFO	TIME
2 mi. @ 3 mph	$T = \frac{D}{R} = \frac{2}{3} = \frac{40}{60}$
3 mi. @ 5 mph	$T = \frac{D}{R} = \frac{3}{5} = \frac{36}{60}$
7 mi. @ 4 mph	$T = \frac{D}{R} = \frac{7}{4} = \frac{105}{60}$
	Total time = $\frac{105}{60}$ hr. = 1.75 hr.

Now, you can use the total time and total distance in the formula $D = RT$. Since you want R, you can rearrange this formula to $R = D \div T$. Thus, you have $R = D \div T = 12 \div 1.75 \approx 6.86$ mph.

Practice Test 2

This second practice test will give you another chance to measure your skills. By this time, you should see real progress in your math abilities.

1. What is the mode of the following numbers?
12, 9, 8, 7, 7, 2, 9, 5, 7
a. 5
b. 7
c. 8
d. 9

Use the chart below as a reference for questions 2 through 3:

METRIC UNITS TO CUSTOMARY UNITS CONVERSIONS
1 cm. = .39 in.
1 m. = 1.1 yd.
1 km. = .6 mi.

2. 3.5 ft. is equivalent to approximately how many meters?
a. 4 m.
b. 3.85 m.
c. 3.18 m.
d. 18 m.

3. 5 yd. 2 ft. is equivalent to approximately how many centimeters?
a. 523 cm.
b. 79.56 cm.
c. 52.3 cm.
d. 6.63 cm.

4. Select the answer choice that best completes the sequence below.
VAB WCD XEF ______ ZIJ
a. AKL
b. UHG
c. YGH
d. GHW

5. 20% of what number equals 40% of 120?
a. 48
b. 96
c. 200
d. 240

6. The ratio of multimedia designers to graphic designers at a production house is 2:1. If the combined number of multimedia designers and graphic designers is 180, and $\frac{1}{2}$ of the multimedia designers are women, how many women multimedia designers are there?

a. 60
b. 80
c. 90
d. 120

7. If a map drawn to scale shows 5.2 cm between two points, and the scale is 1 cm. = 1.5 km., how far away are the 2 points in meters?

a. 7.8
b. 780
c. 7,800
d. 78,000

8. Use ($F = \frac{9}{5}C + 32$) to convert 113° F into the equivalent Celsius temperature.

a. 38°
b. 45°
c. 54°
d. 63°

9. Damian earns a semimonthly salary of $2,300. What is his yearly salary?

a. $55,200
b. $34,000
c. $27,600
d. $24,000

10. It took Amanda 45 minutes to jog 3 miles at a constant rate. Find her rate in mph.

a. 3 mph
b. 4 mph
c. 10 mph
d. 15 mph

11. What percent of $\frac{1}{8}$ is $\frac{1}{32}$?

a. 35%
b. 30%
c. 20%
d. 25%

12. Nicole bought Blue Diamond stock at $15 per share. After 6 months, the stock is worth $20 per share. This represents a percent increase of

a. 25%
b. 30%
c. $33\frac{1}{3}$%
d. 75%

13. One construction job can be completed by 15 workers in 8 days. How many days would it take 20 workers to complete the job?

a. 4 days
b. 6 days
c. 8 days
d. 10 days

14. 3 pieces of wood measure 8 yd. 2 ft. 1 in., 6 yd. 1 ft. 9 in., and 3 yd. 1 ft. 7 in. in length. When these boards are laid end to end, what is their combined length?

a. 18 yd. 17 in.
b. 18 yd. 5 ft.
c. 18 yd. 2 ft. 5 in.
d. 18 yd. 5 in.

15. What percent of $\frac{3}{16}$ is $\frac{1}{64}$?

a. 5%
b. $8\frac{1}{3}$%
c. 33%
d. 80 %

1,200 new nursing students were asked to complete a survey in which they were asked which type of nursing they would like to pursue. The data was used to make the pie chart below. Use this information to answer questions 16 through 18 below:

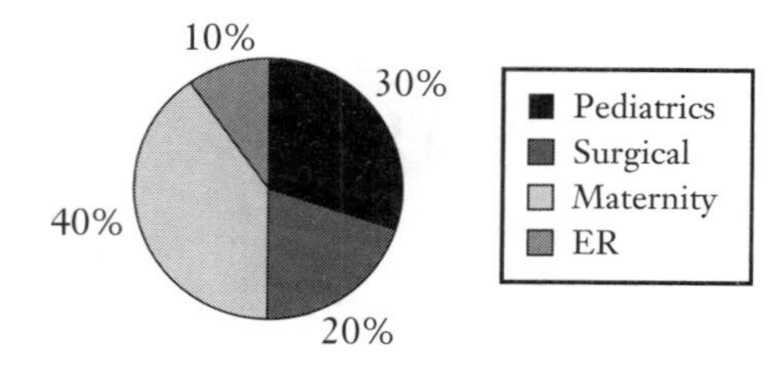

16. How many nursing students would like to pursue pediatrics?

a. 360
b. 400
c. 600
d. 800

17. Half of the nurses who indicated that they would like to pursue surgical nursing also noted that they would like to transfer to a sister school across town. How many students indicated that they would like to make such a transfer?

a. 240 students
b. 120 students
c. 60 students
d. 10 students

18. If the same color scheme is used, which of the following bar graphs could represent the same data as the pie chart?

a.

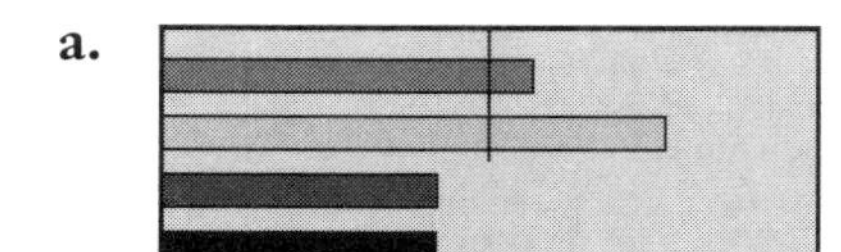

b.

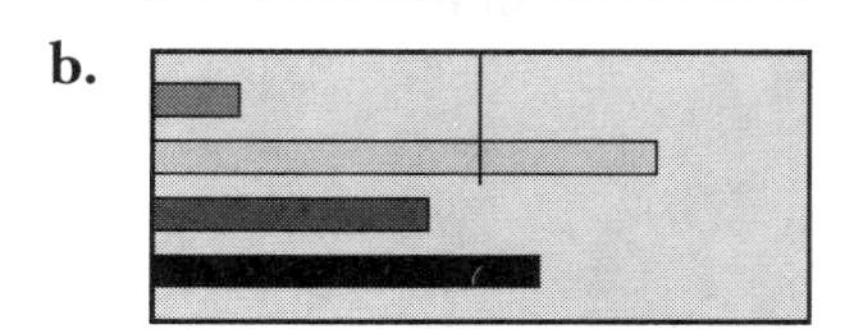

c.

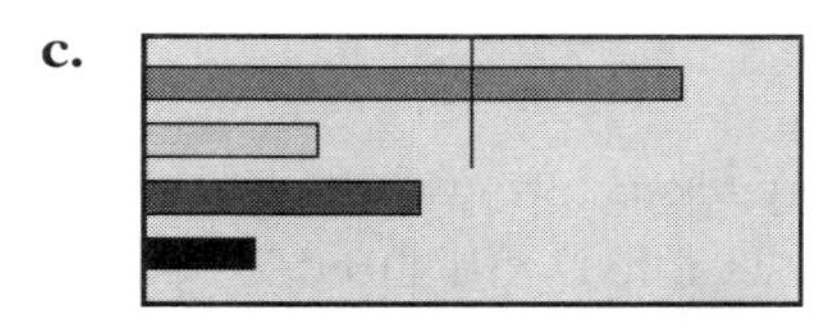

d.

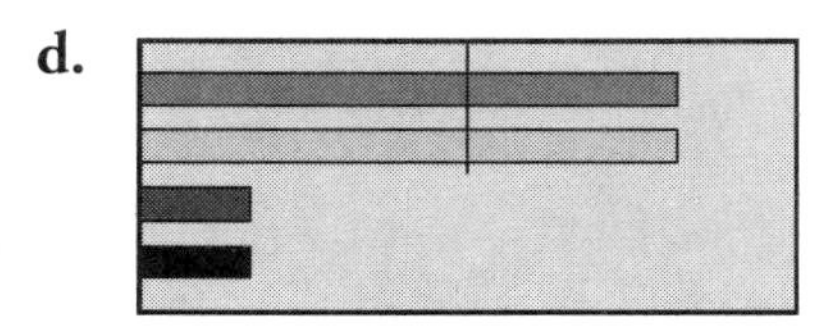

19. $(8^5 \times 3^4) \div (8^3 \times 3^2)$ is equivalent to

a. 576
b. 420
c. 376
d. 256

20. Pipe A leads into a tank and Pipe B drains the tank. Pipe A can fill the entire tank in 1 hour. Pipe B can drain the entire tank in 45 minutes. At a certain point in time, the valves leading to both pipes are shut and the tank is $\frac{1}{2}$ full. If both valves are opened simultaneously, how long will it take for the pipe to drain?

a. $\frac{1}{2}$ hr.
b. 1 hr.
c. $1\frac{1}{2}$ hr.
d. $1\frac{3}{4}$ hr.

21. Select the answer choice that best completes the sequence below.

B_2CD ______ BCD_4 B_5CD BC_6D

a. B_2C_2D
b. BC_3D
c. B_2C_3D
d. BCD_7

22. Select the answer choice that best completes the sequence below.

|┌⊓|□⬠□|⊔ ___ ___

a. □⬠
b. └□
c. |⊓
d. ┘|

23. The reduced price of a computer is $1,250 after a 20% discount is applied. The original price was then

a. $250.
b. $1,000.
c. $1,562.50.
d. $6,250.

24. Three cylindrical solids with $r = \sqrt{7}$m. and $h = 1$ m. are packed into a rectangular crate with $l = 10$ m., $w = 9$ m., and $h = 1.2$ m. The empty space will be filled with shredded paper. What volume will the shredded paper occupy?

a. 86m^2
b. $66\pi\text{m}^2$
c. $42\pi\text{m}^3$
d. 42m^3

25. External hard drives cost $280 each. When more than 30 drives are purchased a 10% discount is applied to each drive's cost. How much money will 40 drives cost (excluding tax)?

a. $7,000
b. $8,200
c. $10,080
d. $11,200

26. Select the answer choice that best completes the sequence below.

BOC COB DOE EOD ______

a. FOG
b. DOG
c. DOF
d. FOE

27. Which of the following rope lengths is longest? (1 cm. = 0.39 inches)

a. 1 meter
b. 1 yard
c. 32 inches
d. 85 centimeters

28. $\frac{2}{5}\%$ =

a. $\frac{1}{250}$
b. .4
c. $\frac{1}{25}$
d. .04

29. A box contains 23 iron washers, 15 steel washers, and 32 aluminum washers. If a washer is chosen at random, what is the probability that a steel washer will be chosen?

a. $\frac{3}{14}$
b. $\frac{23}{70}$
c. $\frac{32}{70}$
d. $\frac{15}{7}$

30. If the volume of a cube is 27 cubic centimeters, what is its surface area?

a. 3 cm^2
b. 6 cm^2
c. 9 cm^2
d. 54 cm^2

Use the chart below to answer question 31 through 33. This graph shows the number of inches of rain for 5 towns in Suffolk County during spring 2002.

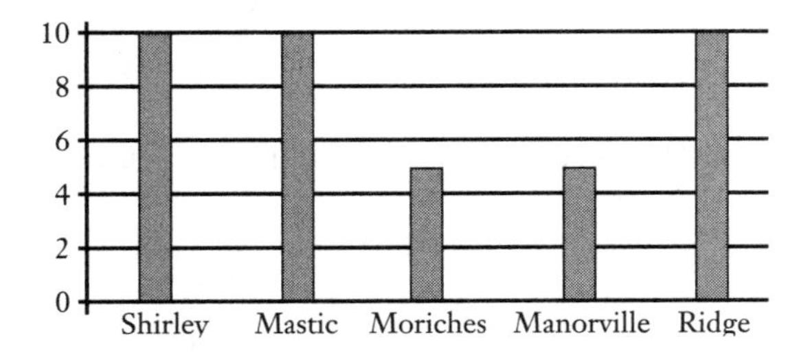

31. What was the median number of inches for the 5 towns?

a. 5
b. 8
c. 9
d. 10

32. What was the mode?

a. 5
b. 8
c. 9
d. 10

33. What was the average number of inches for the season shown?

a. 5
b. 8
c. 9
d. 10

34. When expressed as a percent, $\frac{9}{17}$ is most accurately approximated as

a. .0053%
b. 45.2 %
c. 50%
d. 52.9%

35. The length of a rectangle is equal to 3 inches more than twice the width. If the width is 2 inches, what is the area of the rectangle?

a. 7 square inches
b. 14 square inches
c. 18 square inches
d. 21 square inches

36. Kira's register contains 10 twenty-dollar bills, 3 five-dollar bills, 98 one-dollar bills, 88 quarters, 52 dimes, 200 nickels, and 125 pennies. How much money is in the register?

a. $351.45
b. $351.20
c. $350
d. $345.51

37. Select the answer choice that best completes the sequence below.
DEF DEF_2 DE_2F_2 ______ $D_2E_2F_3$

a. DEF_3
b. D_3EF_3
c. D_2E_3F
d. $D_2E_2F_2$

38. Hannah's yard is square. A light is placed in the center of her yard. The light shines a radius of 10 feet on her yard, which is 20 feet on each side. How much of the yard, in square feet, is NOT lit by the light?

a. 400π
b. $40 - 10\pi$
c. $400 - 10\pi$
d. $400 - 100\pi$

39. Chris drove for 100 miles. During the first 45 miles, he drove at a rate of 75 mph. During the next 45 miles, he drove at a rate of 50 mph. For the last 10 miles, he drove at a rate of 25 mph. What was his approximate average rate for the whole trip?

a. 40 mph
b. 53 mph
c. 55 mph
d. 60 mph

40. What is the area of the shaded figure inside the rectangle?

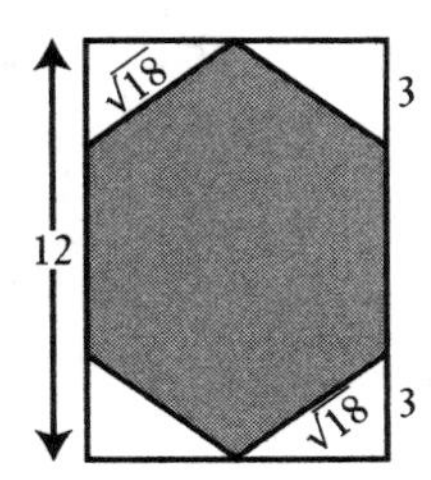

a. 18 units^2
b. 36 units^2
c. 54 units^2
d. 60 units^2

ANSWERS

1. **b.** To find the mode, see which number occurs the most: 12, 9, 8, **7**, **7**, 2, 9, 5, **7**. Thus, 7 is the mode.
2. **c.** You should know that 3 ft. = 1 yd. and the chart tells you that 1 m. = 1.1 yd. Thus, you can create conversion factors that let you cross-off *feet* and end up with *meters*: $3.5 \text{ ft.} \times \frac{1 \text{ yd.}}{3 \text{ ft.}} \times \frac{1 \text{ m.}}{1.1 \text{ yd.}} \approx 3.18 \text{ m.}$
3. **a.** 5 yd. = 15 ft., so 5 yd. 2 ft. = 17 ft. Next, using the fact that 1 ft. = 12 in. and 1 cm. = .39 in., you can create conversion factors that let you cross-off *feet* and end up with *cm.*: $17 \text{ ft.} \times \frac{12 \text{ in.}}{1 \text{ ft.}} \times \frac{1 \text{ cm.}}{.39 \text{ in.}} \approx 523 \text{ cm.}$
4. **c.** The first term of each triplet represents the alphabet in sequence: V → W → X → **Y** → Z. Thus, the first letter of the missing triplet is Y. The second and third letters of the triplets follow the pattern of skipping one letter. Thus, the second term of the missing triplet will be: A → C → E → **G** → I. And the third term of the missing triplet will be: B → D → F → **H** → I. Therefore, the answer is YGH.
5. **d.** "20% of what number equals 40% of 120?" can be written mathematically as $.20 \times x = .40 \times 120$. Dividing both sides by .20 yields:

   $$x = \frac{(.40)(120)}{.20} = 240.$$

6. **a.** You are told that the ratio of multimedia designers to graphic designers at a production house is 2:1. Thus, $\frac{2}{3}$ of the 180 total must be multimedia designers. $\frac{2}{3}$ of 180 = $\frac{2}{3} \times 180 = 120$ multimedia designers. Half of these are woman, so there are 60 women multimedia designers.
7. **c.** First use a proportion to get the real life value: $\frac{1 \text{ cm.}}{1.5 \text{ km.}} = \frac{5.2 \text{ cm.}}{x \text{ km.}}$; $x = 1.5 \times 5.2 = 7.8$ km. Next, convert kilometers to meters by multiplying by $\frac{1{,}000 \text{ m.}}{1 \text{ km.}}$: $7.8 \text{ km.} \times \frac{1{,}000 \text{ m.}}{1 \text{ km.}} = 7{,}800 \text{ m.}$
8. **b.** Substitute 113 for F in the given equation. Thus, ($F = \frac{9}{5}C + 32$) becomes $113 = \frac{9}{5}C + 32$; $113 - 32 = \frac{9}{5}C$; $81 = \frac{9}{5}C$; $81 \times \frac{5}{9} = C$; $9 \times 5 = C$; $C = 45$ degrees.
9. **a.** Recall that semimonthly means twice a month. This means he makes 2 × \$2,300 = \$4,600 per month. Multiply by 12 months per year: $\frac{12 \text{ mo.}}{\text{yr.}} \times \frac{\$4{,}600}{\text{mo.}} = \$55{,}200$ a year.
10. **b.** First, you should rearrange $D = RT$ into $R = \frac{D}{T}$. Substitute the given values into the formula. Here, R = 45 min. = $\frac{3}{4}$ hour, and D = 3 mi. Thus, $R = \frac{D}{T}$ becomes R = 3 mi. ÷ $\frac{3}{4}$ hr. = 4 mph.
11. **d.** The question "What percent of $\frac{1}{8}$ is $\frac{1}{32}$?" can be written mathematically as $\frac{?}{100} \times \frac{1}{8} = \frac{1}{32}$. Recall that *what percent* is $\frac{x}{100}$, *of* means ×, and *is* means =. Solving, you get $\frac{x}{800} = \frac{1}{32}$; $x = \frac{800}{32} = 25\%$.
12. **c.** The Blue Diamond stock rose from \$15 to \$20. This is a difference of \$20 – \$15 = \$5. When compared with the original \$15, $\frac{5}{15} = \frac{x}{100}$; $x = \frac{500}{15} = 33\frac{1}{3}\%$.
13. **b.** If it takes 15 workers 8 days to complete a job, it would take 1 worker 15 × 8 = 120 days. It would take 20 workers 120 ÷ 20 = 6 days.

14. c. First, line up and add all of the units:

8 yd. 2 ft. 1 in.
6 yd. 1 ft. 9 in.
+ 3 yd. 1 ft. 7 in.
17 yd. 4 ft. 17 in.

Next, note that 12 in. = 1 ft., so 17 yd. 4 ft. 17 in. is the same as 17. yd 5 ft. 5 in. Next, note that 3 ft. = 1 yd., so you can rewrite the length as 18 yd. 2 ft. 5 in.

15. b. Recall that "What percent" can be expressed as $\frac{x}{100}$. The question "What percent of $\frac{3}{16}$ is $\frac{1}{64}$?" can be expressed as: $\frac{x}{100} \times \frac{3}{16} = \frac{1}{64}$; $\frac{3 \times x}{1,600} = \frac{1}{64}$; $3 \times x = 25$; $x = \frac{25}{3} = 8\frac{1}{3}\%$.

16. a. 30% (black sector) of the 1,200 nursing students indicated that they would like to pursue pediatrics. .30 × 1,200 = 360 students.

17. b. 20% (dark gray) of the nursing students chose surgical nursing. Half of these want to transfer to the sister school, so that is 10%. 10% of 1,200 = .10 × 1,200 = 120 students.

18. b. If the same color scheme is used (as stated), then in decreasing size order, the bars should be: white, black, dark gray, light gray. Only choice **b** has bars that match this description.

19. a. You can apply the rules of exponents to the terms that have the same bases. Thus, $(8^5 \times 3^4) \div (8^3 \times 3^2)$ is equivalent to $8^{5-3} \times 3^{4-2} = 8^2 \times 3^2 = 64 \times 9 = 576$. Recall that when multiplying and/or dividing exponential numbers, those exponents to numbers with the same base value (i.e. 8^5, 8^3, or 3^4, 3^2) can be either added or subtracted depending on the operation asked to be performed (multiplication → add exponents, division → subtract exponents).

20. c. First, convert the hour into minutes. 1 hour = 60 minutes, so Pipe A fills $\frac{1}{60}$ of the tank every minute. Pipe B empties $\frac{1}{45}$ of the tank per minute. This means the net effect—every minute—is $\frac{1}{45} - \frac{1}{60} = \frac{4}{180} - \frac{3}{180} = \frac{1}{180}$ of the tank is drained. If $\frac{1}{2}$ of the tank is initially full, this equals $\frac{90}{180}$ full. It will take 90 minutes for the $\frac{90}{180}$ to drain out (at a rate of $\frac{1}{180}$ per minute). 90 min. = $1\frac{1}{2}$ hr.

21. b. Notice that the number grows by 1 and moves to the letter on the right of its current position: B_2CD **BC_3D** BCD_4 B_5CD BC_6D. Thus, the missing term is BC_3D.

22. d. Note that the number of line segments increases and then decreases by one: 1 → 2 → 3 → 4 → 5 → 4 → 3. Thus, the next 2 members of the series will have two sides and then one side.

23. c. If a 20% deduction was applied, then $1,250 represents 80% of the original cost. This question is really asking: "80% of what is $1,250?" This can be written mathematically as $.80 \times x = 1,250$; $x = \frac{1,250}{.80} = \$1,562.50$.

24. d. The formula for a cylinder is $V = \pi r^2 h$. If you use $\pi \approx \frac{22}{7}$, and substitute the given values into this formula, you have: $V = \frac{22}{7} \times (\sqrt{7})^2 \times 1 = \frac{22}{7} \times 7 = 22\text{m}^3$. Three such cylinders will occupy a volume of $3 \times 22\text{m}^3 = 66\text{m}^3$ inside the rectangular crate. The volume of the crate is $lwh = 10 \times 9 \times 1.2 = 108\ \text{m}^3$. The empty space (to be filled with shredded paper) is $108\ \text{m}^3 - 66\ \text{m}^3 = 42\ \text{m}^3$.

25. c. Since more than 40 drives are being purchased, use the discounted price. Take 10% ($28) off the cost of each drive. So, instead of costing $280 each, the drives will be $280 – $28 = $252 each. Next, multiply 40 drives by the price of each drive: 40 × 252 = $10,080.

26. a. The first term progresses from B ➔ C ➔ D ➔ E, so the last triplet will begin with F. Note that the second term is always O. Every other triplet is the inverse of the triplet before it. So, the third letter of the last triplet, like its predecessors, is the next letter of the alphabet after F.

27. a. In order to compare the choices, convert them all into inches:

a. 1 m. = 100 cm. = 100 cm. × $\frac{.39 \text{ in.}}{\text{cm.}}$ = 39 in.

b. 1 yd. = 36 in.

c. 32 in.

d. 85 cm. is less than 1 m. (choice **a**) so you need not waste time converting this choice to inches. Thus, choice **a**, 39 inches, is the longest.

28. a. This can be solved by simply equating the percent to its equivalent fractional form(s): $\frac{2}{5}\%$ = .4 % = .004 = $\frac{4}{1,000}$ = $\frac{1}{250}$.

29. a. First, note that all the washers together equal: 23 + 15 + 32 = 70. There are 15 steel washers, so the chance of pulling a steel washer is 15 out of 70: $\frac{15}{70} = \frac{3}{14}$.

30. d. The volume formula for a cube is $V = s^3$, so here $s^3 = 27$ and $s = 3$ cm. The surface area of one face is $s^2 = 3^2 = 9$ cm². Since there are six faces, the total surface area is 6 × 9 cm² = 54 cm².

31. d. First, list the numbers in order. The middle number will be the median: 5, 5, **10**, 10, 10

32. d. To find the mode, select the number that occurs the most:

10, 10, 5, 5, **10**

10 occurs three times and is the mode.

33. b. First, add up all the values: 10 + 10 + 5 + 5 + 10 = 40. Next divide by 5 (the number of values): 40 ÷ 5 = 8 inches.

34. d. First, convert $\frac{9}{17}$ to a decimal: 9 ÷ 17 ≈ .529. Next, to express this value as a percent, just move the decimal point over 2 places to the right ≈ 52.9%.

35. b. "The length of a rectangle is equal to 3 inches more than twice the width," can be expressed mathematically as $l = 2w + 3$. We know $w = 2$, so $l = (2)(2) + 3 = 7$. The area is then $A = lw$ = 7 × 2 = 14 square inches.

36. a. First, multiply the amount of coins (or bills) by the value of the coin (or bill):

10 twenty-dollar bills = 10 × $20 = $200

3 five-dollar bills = 3 × $5 = $15

98 one-dollar bills = 98 × $1 = $98

88 quarters = 88 × $.25 = $22

52 dimes = 52 × $.10 = $5.20

200 nickels = 200 × $.05 = $10

125 pennies = 125 × $.01 = $1.25

Next, add up all the money: $200 + $15 + $98 + $22 + $5.20 + $10 + $1.25 = $351.45.

37. d. The letters remain the same: DEF. The numbers change as follows (a dash, such as "–" represents no number): – – – ➔ – – 2 ➔ – 2 2 ➔ **2 2 2** ➔ 2 2 3.

38. **d.** The area of the dark yard is the area of her square yard ($A = s^2$) minus the circle of light around the lamp ($A = \pi r^2$).

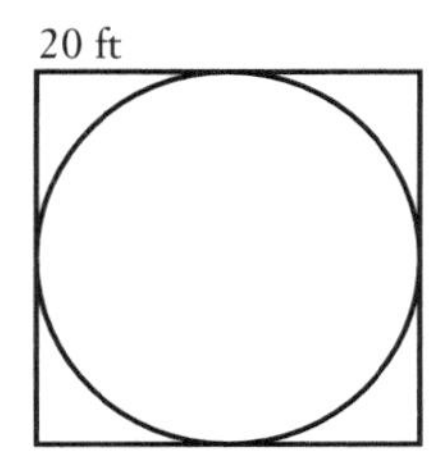

Thus, the dark area = $20^2 - (\pi \times 10^2)$, or $400 - 100\pi$.

39. **b.** To find the average rate, you must use $D = RT$ with the total distance and the total time as D and T respectively. You are given the total distance of 100 miles. You need the total time. This can be found by using the information in the question. The formula $D= RT$ can be rewritten as $T = \frac{D}{R}$. Making a chart for yourself will help you stay organized:

INFO	TIME
@75 mph	$T = \frac{D}{R} = \frac{45}{75} = \frac{90}{150}$ hr.
@50 mph	$T = \frac{D}{R} = \frac{45}{50} = \frac{135}{150}$ hr.
@25 mph	$T = \frac{D}{R} = \frac{10}{25} = \frac{60}{150}$ hr.

***Note that the least common multiple of 75, 50, and 25 was chosen as the denominator for the times listed above.**

Total time = $\frac{285}{150}$ hr. = 1.9 hr.

Now, you can use the total time and total distance in the formula $D = RT$. Since you want R, you can rearrange this formula to $R = D \div T$. Thus, you have $R = D \div T = 100 \div 1.9 \approx 53$ mph

40. **c.** Each little white triangle in the corner is a tiny right triangle with a hypotenuse of $\sqrt{18}$ and a leg of 3. Use $a^2 + b^2 = c^2$ to find the other leg: $3^2 + b^2 = (\sqrt{18})^2$; $9 + b^2 = 18$; $b^2 = 9$; $b = 3$. Thus, the width of the rectangle is $3 + 3 = 6$ units. The area of the entire rectangle is $lw = 12 \times 6 = 72$ units^2. To find the area of the shaded region, you must subtract out the area of the 4 tiny triangles. Each triangle has an area equal to $\frac{1}{2}bh = \frac{1}{2} \times 3 \times 3 = 4.5$ units^2, so the 4 triangles take up $4 \times 4.5 = 18$ units^2. Subtract this amount from the area of the rectangle to find the area of the shaded region: $72 - 18 = 54$ units^2.

Glossary of Math Terms

Area: a measure of the space inside a two-dimensional figure. Area is expressed in square units.

Arithmetic series: a series which progresses by adding (or subtracting) a constant number to each term.

Circumference: the distance around a circle.

Compounded annually: interest is paid each year.

Compounded daily: interest is paid every day.

Compounded monthly: interest is paid every month.

Compounded quarterly: interest is paid four times a year.

Compounded semi-annually: interest is paid two times per year.

Constant rate equation: an equation that is used to relate distance, rate, and time when dealing with a constant velocity: $D = RT$.

Denominator: the bottom number in a fraction.

Diameter: any line segment that goes through the center of a circle and has both endpoints on the circle.

Difference: the answer obtained by subtracting.

Geometric series: a series which progresses by multiplying each term by a constant number to get the next term.

Improper fraction: a fraction whose numerator is greater than the number in the denominator, such as $\frac{8}{7}$.

Least Common Denominator: the smallest number that is a multiple of the original denominators present.

Mean: the average of a set of values. To calculate the mean, follow these steps: **Step 1**— Add all the numbers in the list. **Step 2**— Count the number of numbers in the list. **Step 3**— Divide the sum (the result of step 1) by the number (the result of step 2).

Median: the middle number in a group of numbers arranged in sequential order. In a set of numbers, half will be greater than the median and half will be less than the median. To calculate the median, follow these steps: **Step 1**—Put the numbers in sequential order. **Step 2**—The middle number is the median. (If there are two middle numbers, you find the *mean* (or average) of the two middle numbers.)

Mixed Number: a number that is expressed as a whole number with a fraction to the right, such as $1\frac{1}{2}$.

Mode: the number in a set of numbers that occurs most frequently. To find the mode, you just look for numbers that occur more than once and find the one that appears *most* often.

Numerator: the top number in a fraction.

Order of Operations: the order in which operations must be performed. An easy way to remember the Order of Operations is to use the mnemonic **PEMDAS**, where each letter stands for an operation: **P**arentheses: Always calculate the values inside the parentheses first; **E**xponents: Second, calculate exponents (or powers); **M**ultiplication/**D**ivision: Third, perform any multiplications or divisions in order from left to right; **A**ddition/**S**ubtraction: Last, perform any additions or subtractions in order from left to right.

Percent change: when calculating the percent increase or decrease, equate the ratio of the amount of change to the initial value with the ratio of a new value, x, to 100. The general proportion to use is: $\frac{\text{change}}{\text{initial}} = \frac{x}{100}$.

Percent error: is found by converting the ratio between the calculated value and the actual value to a value out of 100: $\frac{\text{difference in values}}{\text{actual values}} = \frac{x}{100}$.

Percent: a ratio that expresses a value as per 100 parts. For example 30% is equivalent to 30 per 100, or $\frac{30}{100}$. You can express a percent as a fraction by placing the number before the percent symbol over the number 100. You can express a percent as a decimal by moving the current decimal point 2 places to the left.

Perimeter: the distance around a two-dimensional geometric figure.

Prime numbers: numbers that have only 2 factors, the number 1 and itself.

Product: the answer obtained by multiplying.

Proper fraction: a fraction where the number in the numerator is less than the number in the denominator, such as $\frac{1}{2}$.

Proportion: a pair of 2 equivalent ratios in the form $\frac{a}{b} = \frac{c}{d}$.

Quotient: the answer obtained by dividing.

Radius: any line that begins at the center of a circle and ends on a point on the circle.

Ratio: a comparison of 2 or more numbers.

Reciprocal: the multiplicative inverse of a number, for example, the reciprocal of $\frac{4}{5}$ is $\frac{5}{4}$.

Simple Interest: interest is calculated with the formula $I = PRT$. The amount of money deposited is called the principal, P. The annual interest rate is represented by R, and T represents the time in years.

Sum: the answer obtained by adding.

Symbol series: a visual series based on the relationship between images.

The Associative Law: this property applies to grouping of addition or multiplication equations and expressions. It can be represented as $a + (b + c) = (a + b) + c$ or $a \times (b \times c) = (a \times b) \times c$. For example, $10 + (12 + 14) = (10 + 12) + 14$.

The Commutative Law: this property applies for addition and multiplication and can be represented as $a + b = b + a$ or $a \times b = b \times a$. For example, $2 + 3 = 3 + 2$ and $4 \times 2 = 2 \times 4$ exhibit the Commutative Law.

The Distributive Law: this property applies to multiplication *over* addition and can be represented as $a(b + c) = ab + ac$. For example, $3\ (5 + 7) = 3 \times 5 + 3 \times 7$.

Volume: a measure of the amount of space inside a three-dimensional shape. Volume is expressed in cubic units.